KVARKADABRA

Sašo Dolenc
THE GENIUS WHO NEVER EXISTED
and Other Short Stories
from Science, History and Philosophy

Paperback edition 2018

Translated by Gaja Lužnik
Cover design by Arjan Pregl
Copy-editing John Stubbs

Published by:
KVARKADABRA – DRUŠTVO ZA TOLMAČENJE ZNANOSTI
Slovenska cesta 15, 1000 Ljubljana, Slovenia
www.kvarkadabra.net
info@kvarkadabra.net

A CIP catalogue record for this book is available from the Slovenian National and University Library.

001(081)
5/6(081)

DOLENC, Sašo, 1973-
The genius who never existed and other short stories from science, history and philosophy / Sašo Dolenc ; [translated by Gaja Lužnik]. - Ljubljana : Kvarkadabra - društvo za tolmačenje znanosti, 2018

ISBN 978-961-94070-8-0

293778944

Sašo Dolenc

The Genius Who Never Existed

and

other Short Stories
from Science, History and
Philosophy

Contents

Everything should be made as simple as possible, but not simpler.

Albert Einstein

Brain

The man who could not forget

We have all probably wished for better powers of recall at certain points in our lives. When we couldn't remember our credit card pin code at the checkout or when some important telephone number stubbornly evaded us, most of us have probably thought the world would be a better place if we didn't have to struggle to remember information. For this reason it is hard to imagine someone with a completely opposite problem: not being able to forget anything.

The most famous case of inescapable memory was Solomon V. Shereshevskii. He was a journalist of Jewish descent, living in Moscow at the beginning of the 20th century. The first person to notice Shereshevskii's unusual abilities was the editor of the newspaper he worked for. The editor never saw him jot anything down at the daily board meetings, despite the fact that the journalists were given very precise instructions on, for example, where and when a meeting would take place. That's why the editor was convinced at first that Shereshevskii was an unreliable employee, but when the young man recited perfectly all the information he been given in the last hour, the editor realized he had discovered a very special talent.

Because Shereshevskii himself didn't realize his memory was so much better than everyone else's, the editor arranged for him to meet with the young, and soon to be very famous Russian neurologist, Alexander R. Luria. After their first meeting, Luria was fascinated by the young journalist's unusual disorder and he continued to see Shereshevskii regularly over the next thirty years, exchanging letters and meticulously noting his curious troubles and abilities.

The disillusioned Tatars

Luria was the disciple of the famous Lev Vygotsky, the founder of developmental psychology, who died prematurely in 1934 from a fatal attack of tuberculosis. Vygotsky studied many subjects: one of his pursuits involved linguistic and cultural influences on child development and children's perceptions of the world. To scientifically prove his hypotheses, he sent Luria to the far east of Russia to study illiterate Tatars' sense of optical illusions. A typical test set involved comparing simple drawings of patterns, one of which seems larger than the other, while both are in fact the same size.

Luria was so excited about the results that he couldn't bear the long trans-Siberian train ride home without telegraphing Vygotsky in advance, reporting: "Tatars have no illusions." This resulted in his immediate arrest. In those Stalinist times, sending out telegrams referring to the alleged "collective illusions" could cost you your life. Luria later exchanged developmental psychology for neurology and worked as an army brain surgeon.

Unlimited memory capacity

In addition to his scientific papers, Luria wrote an interesting book in 1965, intended for a wider circle of readers, in which he described his sessions with the patient who couldn't forget. He wrote about his decades of attempting to see the world through Shereshevskii's eyes. The English translation of his book, entitled *The Mind of a Mnemonist: A Little Book About a Vast Memory,* was published in 1968 by Harvard University Press.

The book begins with Luria remembering the first experiments he used to test the memory capacity of "S.", as he refers to Shereshevskii. These involved him reading aloud long numeric sequences which S. would memorize and repeat, regardless of their length. S. was even able to recite the sequences in reverse order.

Luria wanted to find as many measurements as possible of the patient's unusual ability, yet became increasingly frustrated at being unable to assess the full extent of S.'s memory. S.'s power of recall namely extended beyond recently registered sequences: he was able to remember the numbers Luria had given him a week, a month or even years before.

Luria states in his book that S. was able to remember a test after as many as fifteen years:

> During these test sessions S. would sit with his eyes closed, pause, then comment: "Yes, yes ... This was a series you gave me once when we were in your apartment ... You were sitting at the table and I in the rocking chair ... You were wearing a gray suit and you looked at me like this ... Now, then, I can see you saying ..." And with that he would reel off the series precisely as I had given it to him at the earlier session.

The downsides of perfect memory

Since Luria couldn't measure the limits of his patient's memory, he focused on S.'s other problems, especially on his very unique perception of the world. S. suffered from a heavy case of synaesthesia, or the merging of sensory impressions. Stimuli in one of his senses activated all his other senses, which often led to unpleasant sensations. After meeting Lev Vygotsky for instance, he declared his voice to be "crumbly and yellow".

Upon entering a restaurant, a random word on the menu could cause such a mixture of tastes and smells that it made him lose his appetite. Neither could he eat and read the paper at the same time. The actual taste of food merged with the taste triggered by the words and made him lose track of what he was reading.

When he and Luria were leaving the institute one day, Luria asked him if he could remember the way. Apparently he forgot who he was dealing with.

"Come, now," S. said. "How could I possibly forget? After all, here's this fence. It has such a salty taste and feels so rough; furthermore, it has such a sharp, piercing sound..."

Synaesthesia may seem like something that might bring great pleasure; when reading a good poem, for example, each word should trigger a particular smell, taste, colour or sound in the reader. But it doesn't work out that way. Luria observed that S. didn't have much appreciation for the poems of Pushkin and Pasternak. He could hardly experience metaphors, with synaesthesia intervening with the actual words and thus hindering any impressions he could derive from their meaning.

Learning to forget

Interestingly, all the potential mistakes in his memory were always mere faults of perception. Because his memory was based on visual imagery, transforming numbers into pictures, he might misplace one of them in a dark corner of his mind when he looked over certain images. "The same thing happened with the word *egg*. I had put it up against a white wall and it blended in with the background."

For a while S. made his living as a circus performer, displaying his incredible capacity for memorizing unrelated information. It was then that it occurred to him that forgetting sometimes might be a good idea. His main concern wasn't whether he would be able to remember all the numbers written on a blackboard that was then removed from the stage, but whether he would be able to prevent his mental images from previous shows mixing together with the new ones.

He solved the problem by developing a system of erasing and rewriting characters on his own mental blackboard. First he obscured the numbers in his mind's eye, and later covered them with new ones. He would then try to figure out how to wipe clear the surface again before the next show. Eventually he realized his power of autosuggestion was so strong, he could forget by simply wishing something to disappear.

Fakir power

With the help of autosuggestion or by immersing in his memories, S. was even able to control his body functions. His pulse was usually seventy beats a minute. Using nothing but the power of thought, he was able to bring it up to a hundred or down to sixty. When Luria asked him how he did it, S. replied:

> There's nothing unusual about it. I imagine myself trying to catch a train that just took off from the station. I need to catch at least the last carriage to make it. Is it unusual then for my blood pressure to rise? Then I imagine myself lying on a bed perfectly still, trying to fall asleep… I start to doze off … my breathing grows calm, my heart beat slows down …

But that wasn't all. By the power of thought alone, he was able to raise his body temperature in one arm and decrease it in the other. He could use autosuggestion to control pain. When he went to the dentist, he simply imagined that the person sitting in the chair wasn't him but somebody else. "I was merely standing there, watching 'him', who was getting his teeth done. Let 'him' feel the pain… It didn't hurt me, you see, but 'him'. I didn't feel any pain." Like a true scientist, Luria added that this dental experiment was never repeated in one of their sessions.

Willpower is a finite resource

One of the secrets to the success of the controversial adventurer and explorer of Africa, Henry Morton Stanley, was his uncompromising attitude towards daily routine. Every morning when he was in Europe he washed, shaved, dressed and wrote in his diary; and nothing could stop him doing exactly the same when he was in the wilderness of Central Africa on one of his long and extremely challenging expeditions. On one mission through a tropical forest, he lost two thirds of his crew to exhaustion, hunger, disease and accidents, and he was the only European in the team who reached the destination after almost a thousand days of trekking through the rain forest.

According to psychologist Roy F. Baumeister and his book *Willpower: Rediscovering Our Greatest Strength* (Penguin, 2011), Stanley's adherence to daily routine despite difficult circumstances was essential to his survival. Because he was aware of the fact that his willpower was a limited resource, by keeping religiously to his normal schedule he saved precious decision-making energy for emergency situations. Even though the sight of him shaving in the middle of the African jungle must have been extremely strange, it gave him the strength to channel his willpower to where it was most needed and leave everything else to habits that his mind could follow on "autopilot".

What Stanley seemed to know intuitively in the nineteenth century has later been confirmed by modern experiments. No matter how hard we try, we only have limited reserves of willpower that slowly run out as we go along. Just like a muscle, our self-control gets worn out with excessive use. What we also know today is that we have a single source of willpower for all the different things we do. If we use it up on one activity, there won't be anything left for others. The mental effort we make to overcome physical pain, for instance, draws on a single reservoir of willpower, which means less is available for social or emotional self-control or for realizing our goals.

Because willpower is a finite resource, it's good to internalize our long-term tasks and turn them into habits. This way, performing those tasks won't use up the limited supplies of active control we have over our behavior. Like the adventurer Stanley, successful writers use routine to write a certain number of pages each day. Despite being merely a mechanical habit, it gives them the freedom and energy to perform other creative tasks.

Self-control as the basis of civilization

Research has shown that people spend a quarter of their waking time resisting temptation. The most common of these temptations proved to be food, followed by sleeping and switching from work to resting or doing something fun. On average, people manage to overcome only 50% of the temptations when they occur.

The ability to resist immediate temptation is not only a sign of good upbringing and the bedrock of health and success, but also the cornerstone of our civilization. Early farmers had to have strong willpower to overcome the desire to eat everything they had produced and to store some either for difficult times or as seed for planting the next crop. The most intelligent non-human primates are able to plan barely twenty minutes ahead and cannot consciously make a decision to store food for later unless this is part of a daily routine they perform instinctively.

In 2010 the results of a large research project on children in New Zealand were published. A thousand young people were followed from birth to their thirty-second birthdays. It turned out that those with a greater degree of self-control were healthier, more successful and less likely to end up in prison. The power of self-control proved a reliable indicator of success in school.

Another piece of research on convicted criminals concurred by proving that those with greater self-control were less likely to reoffend once they'd served their sentences.

Banning certain words from public utterance may seem an absurd social convention, but it could be understood as a sign that we are able to control and hold back our impulsive reactions. It turned out that children who are taught correct bodily posture at school are also more capable of controlling themselves at other activities.

Research on the different circumstances in which people exert their self-control showed that merely looking at an image of an attractive female reduces the level of self-control in an average man, who therefore looks for instant gratification instead of being rational and waiting for a better option. This effect is of course a good friend of advertising.

Researchers have also described the phenomenon of decision fatigue affecting those who have to make a lot of decisions. Decision-making can often exhaust the supplies of willpower in influential individuals, causing them to engage in reckless behaviour leading often to scandals.

In order to maintain our self-control, we are advised to eat nutrients that release energy slowly, providing a steady supply of energy to our willpower. Poor diets can lead to a vicious circle in which our supply of willpower is quickly used up and therefore unable to stop us eating the entire chocolate bar at once – an act that would require a certain amount of self-control for which we need an additional intake of energy. If we don't get the right amount of nutrients in our body, we can't express our will – since to do that we must first put in some effort and energy.

What is free will?

In recent decades, many academic initiatives have shared the goal of proving that people don't really have free will. But the conclusions these studies have reached often stemmed from a definition of free will that doesn't directly correspond to the meaning of the phrase we use in everyday life. A common academic definition of free will is that we are free when in two cases under exactly the same circumstances we are able firstly to first take one option and later choose the other. That way our decision isn't influenced by any direct external factor and we remain the sole cause for it.

However, exerting one's free will in everyday life doesn't mean doing anything we want to do at any given moment. Animals are very prone to this and it is a completely normal and natural response in all living beings. But humans are able to perform acts which entirely go against their own interests.

People are capable of contemplating their decisions by imagining a number of possible future scenarios. Besides the impulses of the moment that favour a particular choice, we are able to take future circumstances into account when making a decision. That's why a hungry vegetarian won't eat spare-ribs even if he's hungry enough to eat a horse, because he knows he would feel much worse when his guilty conscience awoke after he broke his rule.

People are social creatures by nature, living in communities that abide by certain rules. In order to live together we need to be capable of self-control, or able to restrain our immediate desires. People's ability to live in a cultural society depends on whether individuals are capable of controlling their needs if they clash with the accepted rules of that society.

Cultural rituals are often unnatural and absurd in themselves but their point is to demonstrate the ability of mature community members to restrain their impulses. Different eating cultures, such as vegetarianism, halal and kosher are typical examples of people reining in their natural impulse of hunger by following the rules of their particular custom.

To exert our free will we must be able to follow the rules that we either set ourselves or take on from others. Rules in this sense are the societal, personal or communal norms that might contrast with the immediate needs and desires of an individual. We can be free when we stop merely following our impulses and start following rules. Freedom is therefore the liberty we have of applying the mental brakes that selectively restrain our impulsive reactions.

How to raise a genius

When he studied the biographies of great men celebrated in history for their outstanding achievements in science, arts or sports, the Hungarian developmental psychologist László Polgár observed that the clue to their success lay in the early systematic development of the particular skill that brought them world fame later in life. Polgár wrote a number of articles on his findings, including a book called *Bring up Genius*, where he maintained that geniuses are created through long hours of systematic hard work, rather than born with their talents. In the 1960s his ideas were met with intense scepticism, so he decided to conduct an experiment.

He exchanged a few letters on his unconventional ideas at that time with Klara, a young schoolteacher from Ukraine, who wasn't too thrilled by his approach to child rearing at first, but eventually succumbed to his charms and married him soon afterwards. Together they moved to Budapest, where their first daughter Susan was born on the April 19, 1969, the date of the couple's second anniversary.

The curious chess experiment conducted by the Polgár family

To prove his hypothesis on geniuses being the result of upbringing rather than talent, Polgár needed a great accomplishment, one that would be shocking enough to convince even the most hard-minded of sceptics. After a thorough consideration he decided on chess because the level of excellence is easily measurable in comparison to artistic skills. He was merely an amateur chess player himself, and Klara didn't play at all, so his first step was to systematically analyze the literature on chess playing and teaching. He slowly introduced little Susan to the world of chess through stories and games, so she would develop a strong interest in everything related to chess during the first years of her childhood.

Her playtime consisted of her sitting at a game of chess and she never found studying the strategies and moves of the game burdensome. She had a strong inner motivation that was merely guided by her father, who never forced her to do anything she didn't want to. She later recalled that even at that early age, chess was all fun and games to her.

Susan was not yet five years old when her father took her to her first chess tournament. She wasn't taken seriously at all until she confidently started moving the pieces across the chessboard with her tiny fingers. All her opponents at the Budapest youth tournament were at least two or three times her age and she could barely reach the table with the chessboard, yet she won the tournament without losing a single game, causing a sensation.

In 1974 Lászlo and Klara welcomed their second daughter Sofia, followed by their third child Judit. Susan's younger sisters both showed great interest in chess, striving to become at least as good as their older role model. In the next few decades, all three Polgár sisters accomplished everything that could be accomplished in chess. In 1984 Susan was pronounced the world's best chess player and in 1991 she was the first woman in history to be awarded the title of grandmaster. Unfortunately she came across many obstacles during her career, most notably when in spite of having successfully qualified, she was not allowed to play at the world chess tournament because of her sex. After a long struggle with bureaucracy, the functionaries finally gave in and enabled women to play on equal terms against men. The credit for this change goes entirely to Susan Polgár.

Susan's younger sister Sofia entered the history of chess in 1989 with a series of victories at a tournament in Rome where she beat some of the great names of the chess world. The commentators ranked her games among the greatest chess achievements of all times. In 1991, the fifteen-year-old Judit, the youngest of the Polgár sisters, became the youngest grandmaster in chess history, winning the tournaments in which she competed against such notables as Spassky, Karpov, Kasparov, Topalov and Anand. Today, Judit Polgár is considered the best female chess player of all time. She is the only woman that ever ranked among the top ten chess players in the world.

László Polgár's curious family experiment fascinated the public, but nevertheless failed to reach the desired conclusion. It is still commonly believed today that great human achievements are the results of natural talent rather than systematic hard work. The three sisters are regarded as chess players of born genius despite their father's persistent claims that his daughters' talents developed slowly but surely due to months and years of hard work. He wouldn't be too certain about proclaiming his daughters' talents as natural.

The secret life of virtuoso musicians

In 1991 American psychologist Andres Ericsson carried out extensive research, drawing on all the scientific methodology available, in order to find out how one achieves top results. He took students from the Berlin Music Academy and put them in three different groups according to their past achievements and their professors' input. The first group consisted of top musicians on their way to becoming notable soloists, who were regarded as the most talented. The second group was formed by very good musicians who still lacked a certain something to qualify as great. They were looking at careers in leading world orchestras but were not expected to become prominent soloists. The third group consisted of presumably the least talented musicians who were expected to follow a reputable pedagogical career.

After conducting a number of interviews with the students regarding their life decisions and daily routines, Ericsson concluded that there are not many differences between the three groups. Most of them started to play their instrument when they were eight years old and had decided before the age of fifteen to become professionals. Even though he supposedly wasn't familiar with Polgár's theory and his home experiment, Ericsson surprisingly discovered that the key difference between the three groups of students lay where it was least expected. The best students, gathered in the first group, had spent approximately ten thousand hours practicing their instrument by the age of twenty, which was two thousand hours more than the average amount of practice put in by the second group and six thousand hours more than the average of the third, supposedly least talented group.

It turned out that there was no student who would qualify for the first group without the same amount of practice as the other presumably most talented students, while at the same time there was no student with at least ten thousand hours of practice who would not qualify among the top students. Using a more reliable scientific method, Andres Ericsson reached the same conclusion as László Polgár had before him: only systematic long-term practice separates the best from the average practitioners.

Today it is slowly but surely becoming acknowledged that it takes at least ten thousand hours of work to acquire a complex skill in a science, art or sport before becoming an expert in the field. While analyzing scientists and artists at work, it has been established that an average of ten years of systematic work passes between their first and their greatest accomplishments. While it is difficult to work systematically for more than a thousand hours a year in a single field, this finding is still consistent with the rule that ten thousand hours of practice are needed to become an expert.

Can talent be harmful?

In her renowned work *Mindset: The New Psychology of Success* (Random House, 2006), the psychologist Carol Dweck of Stanford

University supports the claim that people commonly explain their achievements and capacities in two ways. She describes those who believe in natural talent and intelligence as having a fixed mindset, and those who are inclined to the influence of environment and systematic hard work as having a 'growth' mindset.

Carol Dweck proved with numerous studies that a fixed mindset could be counterproductive in regard to talent. Because we tend to learn and progress in knowledge by trying to resolve issues that are slightly more complicated than those we already know how to solve, we are bound to encounter failure. We can acquire a new skill only by trying many times and necessarily failing at most of those times. If we are scared of failure, we won't be able to make progress, since failure is essential to it.

The problem with people who are convinced that natural talent exists is that they mistake failure for a sign that they are losing the talent they have been given. That is why they start to fear failure and their progress eventually grinds to a halt. Those with a growth mindset, on the other hand, who perceive talent merely as motivation for continuous learning and experiencing new things, don't get scared of failure because they recognize it as essential to their growth and development. Those who have a fear of failure will thus never be able to achieve excellence. That is precisely why praising children for their presumed natural talents or intelligence damages their motivation. We can't take credit for talents we have been given; hard work and dedication to success are the qualities that truly deserve our praise.

The wisdom of psychopaths

Kevin Dutton, a psychology professor at Oxford University, conducted some fascinating research a few years ago on psychopathic characteristics in the general population. His subjects were drawn from completely different backgrounds and professions. An on-line survey, completed by 5400 people, provided Dutton with a detailed overview of individuals with psychopathic tendencies who tend to pass unnoticed and function normally within society.

In line with his expectations, it turned out that the so-called "functional psychopaths" (a name for psychopathic individuals with no criminal record) were most likely to be found in the ranks of CEOs, lawyers, TV and radio hosts, salespeople and surgeons; while the least psychopathic professions, according to the survey, included social work, nursing, therapy, arts and crafts, styling and teaching.

One participant in the survey, a lawyer, wrote a long letter to Dutton after completing the questionnaire. He described how he had known since childhood that his view of the world was different to that of most people. Interestingly, he felt those characteristics could do more good than harm and concluded that some psychopathic traits might offer a powerful cure to the hardships of modern lifestyles when applied in moderation.

The sincerity of the e-mail encouraged Dutton to start considering potentially positive traits in psychopaths. After a while, he decided to put his hypothesis that psychopaths might be better equipped for solving certain everyday problems to an empirical test. He subsequently published his findings in his book *The Wisdom of Psychopaths: What Saints, Spies, and Serial Killers Can Teach Us about Success* (Farrar, Straus and Giroux, 2012).

Dropping in for a visit on the "evil elite"

While he was researching "the wisdom of psychopaths", Dutton visited Broadmoor, a maximum-security prison in England. Broadmoor is a psychiatric institution for inmates who are considered unsuitable for normal high-security incarceration. There Dutton talked to both doctors and patients. He was mostly interested in the ideas the worst cases of psychopathy - serial killers and other extreme violent offenders - might come up with when presented with certain everyday situations. Could they suggest smarter and more efficient solutions that would still respect the moral boundaries of functional society?

Thus after thoroughly familiarizing himself with the risks and security precautions, Dutton found himself in a secured unit of a psychiatric hospital, surrounded by some of the worst psychopathic cases in England. His guide assured him it was safe to talk to the prisoners and that aggression in psychopaths is generally premeditated and directed towards achieving a specific goal. The patients were supposedly calm and amiable. The first prisoner he interviewed was busy playing a video game when they were introduced, and referred to his fellow inmates as the "evil elite." He accompanied Dutton in meetings with other patients. The next person Dutton shook hands with prided himself on being the most dangerous person in Broadmoor, but assured Dutton that he need not be scared and promised not to kill him. This might sound like a joke, but Dutton saw it was meant as a completely serious guarantee. Dutton's new acquaintances escorted him to other rooms in the unit, presenting him as the expert who had come from Oxford University to write a book about them.

At first, all the inmates were convinced Dutton was going to portray them as monsters without conscience, as they are usually presented in the media. But when Dutton explained that he was mostly interested in what ordinary citizens could learn from them, the patients became much more willing to cooperate. After some initial ice-breaking, the conversation started opening up. Dutton presented his companions with a number of pre-set scenarios, asking for advice on how to react in the situations described. For instance, he wanted to know how the patients would react if they needed to get rid of a tenant who was refusing to move out of their apartment if they had no legal grounds for asking them to move.

After Dutton explained violence was not an option, one of the psychopaths immediately came up with an original solution. He would knock on the door and introduce himself as a state official, who urgently needed to speak to the owner of the apartment. At first, he would refuse to share any further information with the tenants, thus arousing their curiosity. At the point when they could conceal their interest no longer, he would tell them, in a very secretive tone, that routine toxicity measurements had revealed the presence of a highly poisonous gas in the neighbourhood, and that the degree of pollution around the building made "Chernobyl look like a spa retreat". After hearing this, the tenant probably wouldn't take long to move out.

At this time a psychological profiling test was circulating online, by means of which anyone could supposedly check whether they possessed psychopathic tendencies. This was the scenario: at her mother's funeral, a woman notices a man she doesn't know. She instantly has the feeling that he is her soul mate, the perfect man to become her life partner. But unfortunately she isn't able to get his contact information and fears she might never see him again. In a few days she murders her sister. Why?

If you first thought the crime is motivated by jealousy at the sister having had some sort of contact with the mystery man, you are most likely not a psychopath. If, however, your first thought was something like: "of course, the killer's ideal man will most likely attend the sister's funeral just as he attended the mother's", you are on a good way towards succeeding in one of the professions most suited for psychopaths; according to Internet wisdom, at least.

Dutton tested out this story with real psychopaths at the psychiatric hospital. Contrary to expectations no one at Broadmoor suggested the explanation above. One of the patients even remarked that he might well be crazy, since he was staying at a mental institution, but he wasn't stupid and would never come up with such a plan.

A psychopath's brain

A psychopath is generally defined as a person with a characteristic combination of traits, such as charm, charisma, fearlessness, inconsiderateness, narcissism, cogency and lack of conscience. When making moral judgments, psychopaths usually don't bring emotions into the equation, if they are indeed capable of having any, but remain focused on the final result. They can be extraordinarily attractive and convincing if it helps them achieve their goal. But their charm can quickly transform into rude insensitivity if they assess this has a greater chance of success.

Studies of psychopaths' brains have showed that the amygdala, the cerebral region which coordinates fast intuitive emotional reactions to various situations, is far less active when faced with moral dilemmas than in normal individuals. Psychopaths aren't able to feel physically, as most people do through bodily reactions, if they are doing something wrong. They don't get stomach-wrenching sensations, they don't sweat; they experience none of the other somatic signs which usually indicate that someone is doing something they are not supposed to, or might be hurting another person with their actions. They can only stop themselves by rationally assessing the situation, but that is not always as effective as the fast intuitive reaction most commonly known as conscience.

Dutton noticed that psychopaths have a useful capacity for focusing on the present moment which keeps them from mulling over different scenarios and reliving the past – the "what could have happened if" situations. Like the Eastern masters of meditation, psychopaths are able to brush aside all daydreaming about the alternative possibilities events might follow and focus exclusively on the present, which in turn enables them to remain happy and satisfied. According to Dutton, this power of concentration can be, if moderately applied, extremely useful to anyone who has a hard time dealing with the fast pace of modern society. The same goes for other positive features of psychopaths: their charm and their focus on reaching an ultimate goal.

Nature gave us four kinds of happiness

In 1953, two postdoctoral researchers, Peter Milner and James Olds, worked in the laboratory of the acclaimed Canadian neuropsychologist Donald Hebb. Their research involved surgically implanting electrodes into the brains of rats and observing the animals' responses to an electrical impulse stimulating different areas of their brains. The rats recuperated in a few days and behaved more or less normally, the only difference being that they now had wires "growing" out of their heads.

Discovery of the brain's pleasure centre

With one of these experiments the scientists hoped to study a part of the brain they thought was connected with sleep cycles, but the electrodes in the rats' brains accidentally slipped away from their intended location. Initially the scientists activated the implanted electrodes every time a rat went to stand in a particular corner of the cage, but later modified the mechanism so that the rats pushed the button themselves and thus sent electric impulses to their own brains. This modification in the way the impulses were administered contributed to one of the most dramatic results in the history of behavioral neuroscience.

The rats that were sending electrical impulses to their own brains got caught in a bizarre obsession. They pushed the stimulation button up to seven thousand times an hour. The rats were pushing the button so fanatically they even forgot to eat and drink. Even the nursing females suddenly forgot all about their offspring and kept pushing the button non-stop, twenty-four hours a day. Milner and Olds were quick to realize that this experiment was no longer targeting the brain center that controls the sleep/wake cycle, but had led to the discovery of the pleasure center in the mammalian brain.

The next question naturally was the extent to which this experiment applied to the human brain when stimulated in the same fashion through surgically implanted electrodes. The idea may sound preposterous since this experiment today probably would not be authorized by any ethics committee in the world, but it was indeed carried out only a few decades ago.

American psychiatrist Robert Galbraith Heath attempted treating patients suffering from severe depression and schizophrenia by implanting electrodes deep into their brains. In 1972 he published a curious article on reconstructing the Milner and Olds experiment on a twenty-four-year-old male patient, referred to in the paper as B-19.

The results observed in the patient were as shocking as those seen in the rats. When his brain was electrically stimulated in the newly identified area, patient B-19 experienced strong feelings of euphoria and exaltation. When the stimulus was withdrawn, the patient strongly protested and demanded that the treatment continue. A similar experiment was conducted on a woman suffering from chronic pain symptoms. When the electrodes were inserted, she pushed the stimulation trigger with such zeal that she developed sores on her finger and started neglecting herself and her family.

The chemistry of happiness and satisfaction

Today it is no longer a secret that the source of intimate feelings of joy lies in the secretion of so-called happiness chemicals in the brain. It is less known however, that these happy feelings are caused by four different kinds of chemicals, producing four different kinds of happiness. In all the examples mentioned above, the area of the brain being stimulated was the one responsible for producing dopamine, with similar effects resulting from enhanced secretions of serotonin, oxytocin and endorphins.

Endorphins are nature's painkillers and are used to block the feeling of pain. The mechanism evolved to enable wounded wild animals to run from predators. If our brain could release endorphins all the time, we would be able to run with a sprained ankle, which occasionally happens to high performance athletes when they fail to notice an injury. But this mechanism is only intended to resolve life-threatening situations because masking the pain can easily aggravate the injury and possibly even cause death.

The brain starts releasing endorphins when it receives a signal from pain, either physical or emotional. A slight amount of pain is triggered when we run, causing a release of endorphins which can eventually addict us to running. But there aren't many activities where self-injury is small enough to be overridden by the benefits of endorphin production. Opiates have a similar chemical structure to endorphins and unsurprisingly have similar effects, manifesting strong feelings of euphoria.

Oxytocin is a chemical related to trust. It is triggered when we bond with people around us. Dubbed the "cuddle" chemical, it promotes trust, love and companionship. It is a form of reward sent to us by our brains when we successfully form new trustworthy social bonds. The same chemical is released during orgasm and breastfeeding.

The mammalian brain starts releasing increased amounts of oxytocin immediately after a baby is born and reunited with its mother. This keeps the baby close to a safe family environment. Unlike mammals, reptile species haven't developed this mechanism, which prevents offspring from becoming attached to their parents. In regulating their behavior, reptiles rely solely on the feeling of pain or similar sensations. If for instance they don't get enough sun, they feel pain, which wears off when they get warm again. But even when they succeed in reducing the pain, they can't feel happy because their brains don't possess a chemical structure that would induce happiness. The benefit of this pain-avoiding system is that it functions well on a very low energy intake. Since it requires fewer neurons, it is more energy efficient. Mammals' and humans' brains, however, are much more complex and require greater amounts of energy.

When growing up, the child-mother bond in mammals transforms into an attachment to a larger group, the herd in animals and family in humans, eventually leading to a bond with a nation, a political party, a football club or any other group we identify with. When out of sight with other group members, herd animals typically start releasing cortisol, a chemical triggered by stress and an "antidote" to happiness-inducing chemicals.

Cortisol release signals danger, while the secretion of serotonin indicates the danger has passed. Serotonin gives a green light to normal behavior. It lets us know that whatever we are currently doing is safe and that we can continue doing it.

If people around you pay you respect, their behavior triggers a dose of serotonin in you. This is done spontaneously, even if we are not aware of the reason for their respect. Those mammals that occupy a higher position in their group have higher levels of serotonin and are less prone to aggression; low levels of serotonin being the cause of impulsive and asocial acts. Serotonin has two important functions: it lets us know we are out of a danger zone and that it is safe for us to feed.

Dopamine, the last of the four "happiness chemicals", keeps mammals motivated to exert themselves until they reach their goals. It causes excitement about the expectation of something good coming our way. It signals the anticipation of reward but not the reward itself. The release of dopamine is a sign that we are coming closer to the finish line, while oxytocin and serotonin mark the reward we actually get. Dopamine promotes both learning and pleasure in order to make us remember how pleasure was achieved. The highest doses of dopamine are released when the final reward exceeds our expectations, but at the same time it is precisely in the dopamine system that addictions develop.

If we take love as an example of a complex emotional state in terms of brain chemistry, we can identify several chemical components. Dopamine is released in the state of expectation of future pleasures, oxytocin accompanies the formation of trust and serotonin signals security. And in unrequited love endorphins tune in, which makes our concoction of "happiness" about as perfect as it gets.

The marshmallow experiment

Towards the end of the 1960s Walter Mischel, a professor at Stanford University, observed interesting behaviour in his three daughters, then aged between two and five years old. At the approximate age of four, they were able to resist an immediate temptation of fulfilling their wishes if he promised them a reward would follow later. When for instance they begged him for a chocolate bar at the store, he promised them two chocolate bars if they waited till after lunch. He searched the scientific journals and discovered nothing had yet been written on this topic. Since he couldn't find any reliable information on this particular stage of child development, he decided to study it for himself. He designed an experiment to confirm his hypothesis about four-year-old children being able to defer gratification, using children from the kindergarten of a local university as subjects. The interesting thing about this experiment is that in some ways it is still going on today.

In his experiment, Mischel individually guided the kids into a small room adjoining the kindergarten playroom, where they found a children's table and chairs with a cookie or a small cake waiting on the table. In the original experiment he used marshmallow cakes of a kind particularly popular with children, and the experiment was dubbed "the marshmallow experiment".

When the child sat down at the table, Mischel presented the cake and explained that they could eat it immediately or wait a few more minutes for him to come back with another. He proceeded by leaving the room for fifteen minutes and monitored the child's reaction to the dilemma they were facing through a video camera.

A surprising correlation

The resulting footage (duplicates of the original experiment are available on the internet) is incredibly entertaining. Some of the kids couldn't resist the temptation and gobbled up the marshmallow while the researcher was still in the room or immediately after he had left. Others had major inner conflict written all over their faces. All kids were tempted to get another piece by following the instructions and waiting for the researcher to return, but it certainly wasn't easy.

Mischel found that four-year-olds were capable of waiting an average of seven minutes before they succumbed to their craving. Of course some held out for twenty minutes, and others for less than a minute. But its immediate results weren't what made this experiment enter the history of science.

The research on self-control in pre-school children would by itself be nothing special if Mischel hadn't by chance discovered a correlation to the original marshmallow experiment some years later. He made the discovery while talking to his daughters about their kindergarten friends who had participated in the experiment. They were trying to remember their names, wondering which school they had gone to, what they were doing now and how they are perceived by other children of their age group. It was then that Mischel realized, to his surprise, that there was a strong correlation between the levels of success his daughter's friends had in school and their earlier ability to delay gratification during the experiment.

The kids that couldn't resist the temptation and ate their cake immediately, knowing they could get another if they waited for just a few minutes, generally weren't doing so well in school. At first it seemed a coincidence, since it applied only to his daughter's friends, but Mischel decided to track down other children who had participated in the initial tests and see if there was a similar pattern in the bigger sample.

He got astounding results. From the length of time a four-year-old waited before munching on a marshmallow, Mischel could accurately predict how successful this child would later be in school. Children who had a lower degree of self-control and couldn't wait alone in the room without immediately eating the marshmallow later developed problems in school. Others, who exhibited a greater degree of self-control, had fewer difficulties.

An on-going experiment

Mischel continued monitoring the children's progress. Around the year 1990 he reported significant differences in their SAT scores. The children who had waited for less than a minute before eating their marshmallow had on average 210 points less than their peers who had held out for ten minutes or more.

The kids that participated in the original research are now about 45 years old and Mischel is still monitoring their behavior. He tries to gather as much information on them as possible, from how much they earn, how satisfied they are in life, their health, their level of education, to whether they have had marital problems, weight or obesity issues, etc.

In a recent interview Mischel stated that the differences he observed four decades ago in the kindergarten are still shaping the fate of these individuals. The kids that were able to delay the instant gratification of their cravings now have better jobs, earn more money, are healthier and generally more satisfied with their lives.

Inborn or acquired?

But what does this mean? Can children at the age of four really be divided between those who will become successful and others who will be "losers" unable to control themselves, with nothing but problems awaiting them in life? Mischel insists that the experiment should be studied in great detail before any such conclusions are drawn.

The footage of the experiment shows all the children suffering great torture at having to watch the treat before them. In their minds, they probably suffer a conflict between their desire to eat the cake instantly and a drive to resist temptation and receive a greater subsequent reward.

It is interesting to observe how small children are already capable of coming up with different ways to fight the temptation and not eat the cake. Some of them sniffed around it, others covered up their eyes, others still only licked it so no bite-marks would show. The more inventive sang songs or counted on their fingers to pass time.

While watching the footage, Mischel noticed that the children who managed to hold out longer before eating the dessert developed efficient distraction techniques to divert their attention temporarily to something else. Whether by counting, playing with their fingers, singing, kicking the table, all these techniques enabled them to hold out longer.

For this reason Mischel then started teaching these diversion techniques to kids who were initially unable to control their impulses. He told them to imagine the cake in front of them wasn't a real cake but only a picture of the cake. By following such simple suggestions, the children who couldn't wait at first became increasingly patient. These tricks applied to a concrete task, of course, and couldn't influence a child's future life, but it was clear that practice and learning made a great difference.

Mischel predicts that future research will show to what extent the ability to delay gratification is an inborn or an acquired quality. He predicts that it will most probably turn out to be a combination of both. Some of us may be naturally prone to demand immediate satisfaction while others are innately better at waiting; but both tendencies can be greatly influenced by learning and training.

Thinking, fast and slow

In his recently published book *Thinking, Fast and Slow* (Farrar, Straus and Giroux, 2011), Daniel Kahneman, psychologist, winner of the Nobel Prize in economics and expert on the psychology of judgment and decision-making, describes an interesting experiment conducted not long ago in Israel. The researchers spent ten months studying the work of eight parole judges who in that period of time reviewed 1112 parole applications. Each deliberation lasted for an average of six minutes and the judges approved 35% of the applications.

Among other variables, the judges' deliberations were studied in relation to the time of day their decisions took place, and the conclusions were fascinating. When the researchers drew a graph of the number of approved applications in relation to time, it became clear that the judges were more likely to award parole in cases they heard immediately after taking a meal break. 65% of paroles were awarded after the judges finished their meals, but their leniency towards prisoners soon began to wear down and would almost completely disappear when the next meal.

The researchers looked into all possible interpretations that would explain these results, but in the end, the most obvious explanation proved to be the most likely: when hungry and tired, the judges make what is for them the simpler of the two decisions – denying parole.

The Florida effect and the cross-pollination of ideas and practices

Another experiment described by Kahneman in his new book is one where a group of American students was asked to form a sentence out of four words chosen from a list of five. The students were not aware that half of the group received a list of words that were in one way or another connected to old age, such as: wrinkle, bald, grey, forgetful, Florida, etc. When they finished the first part of their task, the students were immediately sent to an office at the other end of the building to continue with the experiment. What they were not aware of was that this part was actually the real experiment – for the main goal of the researchers was to time how quickly the students would reach the other office. It turned out that the students that formed their sentences with words evoking old age walked more slowly and took more time to reach the other end of the building than the students who were given "age-neutral" word-sets. The researchers named the experiment "The Florida effect".

What is also interesting is that the effect takes place without the participants realizing they have been manipulated. When the students were later questioned about a recurring theme in the choice of words, they insisted that the words in the first part of the experiment had no bearing on their subsequent behavior. Age-related issues were not something they would consciously consider during the experiment, yet still had a profound influence on their actions.

It was later found out that the relationship between ideas and practice is a two-way street. Not only do ideas influence practice, practice serves as a guide to directing ideas. In this case the test group students walked for five minutes at the speed typical of an elderly person. Later, their speed at recognizing certain words was also measured. It was found that the group that walked at a slower pace was faster at recognizing the words evoking old age than the group that walked at an average pace. The connection works both ways: if you consciously or unconsciously think about being old, you will act as though you are old, and if you act as if you are old it will make you unconsciously think about being old. A similar cross-pollination of ideas and behavior had been observed in a number of different cases, which proved very advantageous for marketing strategies, to name but one example.

Another piece of research showed that when something in the room reminded them of money, say a picture of a bank note on the wall, participants in the experiment become more persistent in dealing with complicated tasks. A sign triggering the unconscious idea of money results in greater individualism and unwillingness in participants to work together.

Other interesting experiments include one conducted in a kitchenette at one of the British Universities. Its users kept a penny bank where they collected voluntary contributions for coffee, tea and other necessities that were later available to all users of the kitchen. For a few weeks, the researchers kept systematically changing a picture hanging above a refrigerator. The first week they put up a picture of flowers, but replaced it the next week with a cut-out of a face, showing only a pair of eyes staring directly at anyone who walked into the kitchen. It is hardly surprising that more money was put in the penny bank when the flowers were replaced by the gazing eyes. The sense of being watched automatically makes us better behaved, a truth exploited widely in totalitarian countries where a portrait of the leader hangs in every room.

Fast and slow brain systems

The experiments mentioned represent a small part of a large number of similar studies conducted in recent decades, which invite us to start using in our everyday lives a model of human cognition based on the latest scientific discoveries. We can describe this new model with two separate mechanisms of thinking that coexist in our minds. Kahneman refers to them as System 1 and System 2; the intuitive system and the rational system.

The intuitive system can be described as an unconscious autopilot that guides our actions most of the time, and proves to be far more influential than we are usually ready to admit or realize. It is the author of nearly all of our decisions. The intuitive system functions quickly and automatically, without any apparent effort or us realizing what is going on, while the activation of the rational system requires us to produce directed attention and conscious effort. Both systems are functioning continually, with the rational system usually working at a minimal level since it is prone to wasting a lot of energy. In order to save energy, the rational system is naturally lazy and prefers to not get involved with the decisions of the autopilot, unless it is forced to.

Task distribution between the two systems has proven to be very efficient. The conscious system adopts the advice of the autopilot without any contemplation and without realizing that its decisions have been pre-conceived. The unconscious intuitive system is constantly sending messages to the conscious rational system in the form of signals that we are able to recognize as emotions not connected to a specific situation. These intuitive directions, that are often unconscious, are crucial to our normal functioning. When the intuitive system encounters a problem and is unable to make a decision, help comes from activating the rational system. This usually happens when the autopilot is caught unprepared or hadn't been programmed to deal with a particular event. The most obvious indication that the rational system has switched on is dilation of the pupils.

Another characteristic of the intuitive system is that it can't be turned off. It is always active, whether we like it or not. This is very well demonstrated by the above-mentioned examples, where unconscious behavior guided by the intuitive system caused the majority of unusual results. It's therefore good to have some preventative understanding of when we are being deceived by our autopilot and the kind of situations that are more likely to activate the rational system. Intensely focusing attention on a particular task performed by the rational system can narrow down our perception of the situation. This effect has been exploited for thousands of years by all kinds of swindlers who deliberately divert our attention in order to let their trick pass unnoticed.

What about you?

To bring all this together, a simple experiment can check our rational system's rate of activation. Try to solve the following problem as quickly as possible:

> A bat and a baseball ball cost together 1 dollar and 10 cents. The bat costs 1 dollar more than the ball. How much does the ball cost?

Which price did you think of first? What is the solution offered by your intuition? Ten cents, of course. But unfortunately the solution suggested by your autopilot isn't correct. Even though the result doesn't add up, we intuitively think of this solution as the right one. If you answered '10 cents' it means you haven't activated your rational system at all, otherwise you would check the results by means of mathematical calculation. The experiment has already been conducted on a great number of American students with quite shocking results. More than half of the students at Harvard, MIT and Princeton gave a wrong answer, with worse results at less prestigious universities, where the wrong answers amounted to as much as 80%.

The organ for happiness

What would happen if Ingrid Bergman made a different decision in the famous airport scene at the end of the film *Casablanca*? Would she be less happy with her life if she stayed in Morocco with Humphrey Bogart instead of joining her husband on the flight to Lisbon? Would she regret her decision? According to Daniel Gilbert, a professor of psychology at Harvard and one of the world's most renowned researchers of happiness, any choice she made would ultimately lead her to the same feeling of happiness. In his book *Stumbling on Happiness* (Knopf, Random House, 2006) based on the latest scientific discoveries - to which he himself has greatly contributed - Gilbert gives an original view on a thousand-year-old question that still features prominently in thoughtful musings and philosophical debates: How to be happy?

The brain as the experience simulator

In a few million years, the average size of the human brain has more than doubled from that of our primeval ancestors. That's mostly due to the formation of the frontal lobe. One of the most important functions this newer part of the brain performs is allowing us to rehearse experiences. Like pilots who practice different airborne situations in flight simulators to test their reactions, we each possess a cerebral "simulator" that enables us to run through different situations mentally before they actually take place. This is something no other animal is able to do as well as humans.

Synthesized happiness

According to Gilbert, this part of our brains has the extremely beneficial power of helping us to simulate conditions in which we are happy. He describes the brain synthesizer of happiness as a sort of psychological immune system that allows us to picture the world as a better place than it actually is. Even though we are not conscious of it, we all possess an in-built organ for producing happiness in our heads. We continuously produce happiness as we go through life, yet still we perceive contentment as something that we never get and over which we have no control.

What does all this mean? Are there two kinds of happiness: a natural one that stems from the truly happy events in our lives, and a synthesized happiness – a sort of a defense mechanism that paints an individual's existence as being prettier than it is in real life?

Can we distinguish between natural and artificial happiness?

Natural happiness is therefore a feeling we get when our wishes come true, whereas synthetic happiness is something our brain produces as a consolation prize when we don't get what we want. A common belief in western society is that synthetic happiness is worth less than the natural kind. This might be attributed to our consumer mentality, with our being constantly told we will be happier if we do or buy a certain thing.

This is actually not far from the truth. When we buy a new pair of shoes or a TV set, we feel content. But according to researchers' findings, we would be just as happy if we didn't get what we wanted. It would set off a mechanism for artificially inducing a feeling of happiness and the final outcome would be just the same. The results of the experiments conducted confirm the assumption that both sources of happiness, natural and artificial, feel equally real.

The classic free choice test

Gilbert describes a test that has been known in psychological literature for decades. A randomly chosen person is presented with a choice of six objects and is asked to rank them from the one they like least to the one they would most like to have. They usually do this by looking at pictures of the objects. Then they are offered one of the two objects they put in third and fourth place on their list. Since the two things got an average ranking, the test person has more or less the same attitude towards both: they aren't the best but, at the same time, not the worst choice either.

After some time (it can be a couple of hours or a couple of weeks), the person is asked to rank the same objects again. Generally, the classification changes. The object that the person received at the last time of asking now features one place higher than before, and the object they didn't choose now comes one grade lower.

Gilbert attributes this change to the mental mechanism for synthesizing happiness, which was set off when the first choice was made. When the person decides on an object, the mechanism for artificially synthesizing happiness convinces them they have made the right choice. The possibility they opted for is presented as a much better choice than the one they rejected.

Experiment with amnesic patients

Gilbert carried out the same experiment with people suffering from anterograde amnesia, a condition which stops you from forming new memories. As in the previous experiment, Gilbert asked his subjects to rank a set of objects from one to six, according to preference. Then, as before, he gave them the choice of keeping their third or fourth favorite item, this time promising to send it in the post.

After thirty minutes, when the amnesic patients could no longer remember the event, Gilbert returned to their room and told them they had just met (which the patient of course couldn't remember) and asked them to rank the objects again from one to six. But before they started, he asked them if they remembered which object they chose half an hour ago. Their guess was as random as a throw of the dice, which meant that they (consciously) didn't remember which object was already theirs and would arrive in the mail.

But surprisingly, the amnesic patients, like the control group of healthy individuals, proceeded to rank the object they had chosen earlier higher than before. What happened? Gilbert explains that both groups changed their perception and evaluation of the world after they'd made a choice. The case with amnesiac patients proves that evaluation isn't guided by a consciously premeditated belief that one should praise one's own virtues, but that their scale of priorities had truly changed, since the patients had no information about which object they'd chosen and therefore couldn't have reconsidered their own choice beforehand. What actually changed was their instinct for evaluating the world.

How does it work?

To gain a better insight into the mechanism for synthesizing happiness, the researchers conducted another experiment at Harvard. They invited students to photograph their professors and schoolmates and told them they could develop two of their favorite photographs and have them framed in a large format. Then the researchers told the subjects that they were allowed to keep only one of the photographs, while the other would be sent somewhere far away and they would most probably never see it again. The students had to choose which of the two photographs to keep and which to give away.

Half of the students were offered the possibility of changing their minds over the next few days before making a final choice, while the other half was told that their choice would be final and effective immediately. The researchers monitored both groups' attitudes towards the photographs they kept. The students who had the option of modifying their choice saw the photograph they kept as being only slightly better than the photograph they gave away. Their grade they gave their choice didn't improve even after the deadline had passed. On the other hand, the students who were given no option of changing their minds regarded the photograph they kept as much more valuable than the one they'd given away.

The paradox of free choice and the feeling of happiness

Freedom of choice is a great friend of natural happiness, but an equally great enemy of synthesized happiness. The intention of the psychological immune system is to make people accept situations over which they have no control. The artificial production of happiness works best when we are completely stuck, when we have no possibility of choosing freely. The photography experiment was conducted again with a different group of students who now were allowed to choose the group they would like to work in. Two thirds of the students decided they would rather participate in the group that offered them the chance to change their minds before making a final decision.

From the viewpoint of ultimate satisfaction, this is a strange decision to make, since those in the group which had no option of changing their minds ended up much happier and more satisfied with the choice they had made.

Can animals think?

Influenced by Darwin's discoveries on the evolutionary origins of man, from the early twentieth century the wider public became intrigued by all kinds of new information that might shed some light on the actual mental capacities of our closer and more distant animal relatives. Stories of extraordinary abilities in some specially bred animal species spread quickly among the crowds.

Clever Hans

German gymnasium mathematics teacher and amateur horse trainer Wilhelm von Osten was convinced that his horse Hans was remarkably intelligent. He claimed the horse could do arithmetic and provide correct answers to a number of simple problems. He also supposedly understood German, since he answered verbal questions by stomping his hoof. When his master asked him how much was seven plus four, the horse stomped eleven times. In the same way he was able to tell the time, what the date would be the following Friday, and occasionally performed more complex mathematical operations such as fractions.

Hans's mental capacities aroused great public interest. In order to make sure that Hans's performances involved no cheating, the German board of education appointed a commission of thirteen experts consisting of a veterinarian, a circus manager, a cavalry officer, a number of school teachers, and the director of the Berlin zoological gardens. In September 1904 the commission concluded that no trace of fraud could be found. The story was covered by *The New York Times*.

In 1907, a formal investigation of Hans's incredible abilities was carried out by a doctoral psychology student called Oskar Pfungst, an assistant to the German philosopher and psychologist Carl Stumpf. He tackled the research methodically and soon realized that the horse could do mathematical calculations regardless of who was his questioner or audience, but also that the horse answered correctly only if his questioner knew the answer. Nor could the horse give a correct answer if he couldn't see the questioner.

Through further analysis Pfungst showed that Hans could sense small changes in the manner, posture and facial expression of the questioner at the moment when he reached the right number of stomps. Hans wasn't solving problems, but simply stomping until he noticed a change of behavior in the person questioning him.

It turned out that the horse's master didn't even realize he was guiding the horse's reactions by his bodily movements. These minute unconscious gestures proved to be deeply rooted in human behavior, to the extent that Pfungst wasn't able to control them even after he already knew where the secret of Hans's intelligence lay. He did learn to read body language, though, and used it the same way the horse did, answering as many questions as Hans.

The horse's master Wilhelm von Osten never acknowledged Pfungst's findings and continued to tour Germany, mesmerizing the crowds with his equine prodigy. However the young psychologist's findings did have a profound effect on science, or more precisely on experimental methodology in animal and human research.

In the case of Clever Hans, it became clear that the questioner is likely to suggest the right answers to the person they are interviewing even if they don't intend to do so consciously. This was the beginning of the so-called double-blind method, where neither the tester nor the subject are allowed to know whether, for instance, the pill that is being administered contains an active ingredient or whether it is merely a placebo. Today, all medication needs to go through these tests to prove it is efficient enough to be put on the market.

But despite the risk of the Clever Hans effect, the question about the actual mental capacities of animals still remains unanswered more than a century later.

Alex the talking parrot

After Irene Pepperberg saw a 1976 TV documentary about a chimpanzee that learned to use sign language, she decided not to pursue her doctorate in chemistry at Harvard, but to specialize in the study of animal cognition instead. She wanted to find out what an ordinary parrot could learn with systematic training.

She went to a pet store and asked an employee to pick out one of the young African Greys. She named the parrot Alex, an acronym for Avian Learning Experiment, and began with a study that completely transformed our understanding of what birds can do.

Pepperberg presented her intention of proving birds were able to think and talk when taught to do so in one of her first funding requests, which was refused by the grant reviewer with a comment that he didn't know what exactly she had been smoking when she wrote it. At the time she joined the project, scientists were convinced that only primates had sufficiently complex brains to learn elaborate communication with humans. The main goal of Pepperberg's research was to show that the parrot's learning process didn't merely consist of a set of mechanical reactions, but that it involved actual thinking, similar to that in humans and higher primates.

Pepperberg spent decades systematically teaching Alex to repeat the gestures and voices of a student who served as a model. First the student responded to a question asked by the experiment manager. Alex would repeat her answers and then gain positive reinforcement. The course was intense and lasted for more than eight hours a day every day, and since Alex wasn't always in the mood to learn, he sometimes protested by deliberately saying the wrong thing, even though it was obvious he knew the right answer.

Years of working with a parrot took many by surprise in showing that birds were also capable of fascinating cognitive achievements that topped even those performed by the Clever Hans. Alex learned to pronounce 150 words, used them in conversation and combined them in expressions. He also learned to distinguish among 50 objects, defining their shapes and colors, as well as to use numbers from 1 to 6 actively, unlike his predecessor Hans a century earlier.

An important finding of the research was that a parrot was able to identify a common trait between two objects as well as their respective differences. When asked about the difference between a plastic yellow key and a real metal key he answered, "color" and when he was inquired about their similarities he answered, "shape".

In 2005 Irene Pepperberg, now a professor at Harvard, reported that Alex understood the concept of zero. When she asked him about the differences between two objects, he answered in the same way he did with the keys. But if the objects were identical, he answered there was no difference, which could suggest that he understood the idea of zero or the absence of a difference. Alex's acquired intellectual capacities were supposedly comparable to those of a five-year-old child, while emotionally he was equal to a two-year-old. This estimation by Irene Pepperberg may seem utterly unrealistic, but the longer one watches footage of Alex on YouTube, the more convincing it becomes.

On the September 6, 2007, Alex was found dead in his cage. The previous evening, he and Irene had said goodbye in their usual exchange of a few words. Alex died at the age of thirty-one, reaching only half the usual lifespan of a member of his species. Vets determined that the most probable cause of death was a heart condition, even though the bird had passed his regular health check only two weeks earlier.

There were lengthy reports about Alex's death in all major newspapers, from *The New York Times* to scientific magazines such as Nature. *The Guardian* reported: "America is in mourning. Alex, the African Grey parrot who was smarter than the average US president, has died at the relatively tender age of thirty-one."

Pushing the limits of free will

On the August 1, 1966, twenty-five-year-old Charles Whitman climbed to the top of the Tower of the University of Texas in Austin. Armed with several firearms and a pile of ammunition, he went to the observation deck and started shooting. Before the police brought him down, he shot and wounded a number of random passers-by, including a pregnant woman and some brave individuals who came running to help her.

Soon after "the Texas Tower massacre" that shocked the nation, a search of the young man's home revealed two bodies. Before going on his shooting spree, Whitman had murdered his wife and mother and even found time to sit behind his typewriter and write a suicide note. It reads:

> I don't really understand myself these days. I am supposed to be an average reasonable and intelligent young man. However, lately (I can't recall when it started) I have been a victim of many unusual and irrational thoughts. … It was after much thought that I decided to kill my wife, Kathy. … I love her dearly, and she has been as fine a wife to me as any man could ever hope to have. I cannot rationally pinpoint any specific reason for doing this.

After a killing spree committed by what seemed a harmless family man, the general public couldn't wrap its mind around the fact that a person with an IQ of 138 could lose his mind so completely and suddenly. After an autopsy on Whitman's body was performed, some things became clearer. A tumor was found in his brain, pushing against the amygdala, the control center for emotional reactions, in particular fear and aggression.

It seemed that Whitman had been aware of his actions the entire time, but he still couldn't help himself committing the hideous crimes. In his suicide note he added:

> If my life insurance policy is valid ... donate the rest anonymously to a mental health foundation. Maybe research can prevent further tragedies of this type.

Law and the brain

This story of an ordinary young man who transformed into a mass murderer because of a brain tumor is David Eagleman's prologue to his book *Incognito: The Secret Lives of the Brain* (Pantheon, 2011). Eagleman discusses the relationship between law and research on the human brain. In addition to Whitman's case from the 1960s, he describes other examples from medical and judicial archives.

In the early morning on May 23, 1987, twenty-three-year-old Kenneth Parks from Toronto got out of bed and, without waking up, got into his car and drove twenty kilometers to the home of his in-laws. He broke into the house and stabbed his mother-in-law to death, while his wounded father-in-law managed to escape. Immediately after committing the crime, the "sleeping killer" drove to the police station and stated in confusion: "I think I killed somebody." His bloodstained hands testified to his claims.

During court proceedings it turned out that Kenneth had no motive for his crime and that he wasn't consciously aware of his actions, since he hadn't even been awake as he perpetrated them. With the help of testimony from sleep experts who found strange patterns in his brain activity while sleeping, his defense team managed to convince the court that Kenneth couldn't be held responsible for unpremeditated actions he hadn't been aware of at the time they were committed. On May 25, the jury pronounced him not guilty of murdering his mother-in-law and attempting to murder his father-in-law.

Another case Eagleman mentions in his book is that of a forty-year old man who, according to his wife, suddenly started to develop pedophilic tendencies. He was brought in to be examined by a doctor, who discovered a large brain tumor. After surgery, his sexual preferences went back to normal, but for no longer than 6 months. Again, he was brought to the hospital, where they noticed a tiny part of the tumor had been overlooked and started to grow again. After a second operation, his symptoms never reappeared.

Is freedom possible?

The more we learn about how our brains work, the more insight we gain into strange and inexplicable types of human behavior. But scientific progress keeps pushing the boundaries between so-called free-will decisions and the pathological developments which prevent some people from acting according to their consciences. Eagleman compares today's monitoring of brain activity to an astronaut observing individual countries on Earth from a space ship or satellite orbiting around it. He can tell a lot about the countries, but far less than their inhabitants on the ground.

By referring to a number of medical cases, Eagleman shows that the boundary between an act of free will and the biological background allowing or impairing it greatly depends on scientific development. Due to considerable progress in neuroscience, scientists are better able to identify problems in cerebral mechanisms, as well as disturbances caused by congenital conditions, environmental factors, childhood traumas, chemical influences or injuries. This has led to legislators questioning the credibility of free-will decision-making and, consequently, the basis for sentencing policy. Is freedom limited by unconscious mechanisms that lead us to commit a certain act?

Eagleman questions whether it still makes sense to determine a boundary between biological causes and free will. Even if the causes of criminal behavior remain unknown, one can presume something must be wrong with the criminal, since the criminal act itself already constitutes a symptom, an indicator that something is not as it should be. The only thing that expert court testimonials can tell is whether today's knowledge and technology are able to explain a certain act by observing brain function disturbances, but not whether these disturbances actually exist. Eagleman suggests redirecting the court's attention from deciding whether the criminal is guilty or not to asking what kind of consequences the act could have for the society. The sentence should therefore be oriented towards the future by considering the potential threat a criminal poses to society and the prospects of rehabilitation.

But in practice even estimating the possibility of relapse proves problematic. During an examination of US parole proceedings for sexual transgressions it turned out that the experts taking part weren't able to tell which offender was more likely to reoffend. Their specialist analysis proved no more accurate than the flip of a coin.

This realization was followed by research that included more than 20,000 sex offenders and an especially wide range of data, such as whether the offenders had suffered abuse in childhood, their marital status, whether and for how long they had been in a serious relationship, history of drug or alcohol abuse and so on. The people who took part in the research were then monitored for five years to establish which traits defined the offenders more prone to relapse. The findings later informed the more reliable guidelines and charts that are now used in determining parole.

Another problem is rehabilitation. The proliferation of lobotomies, today a thing of the past, was initiated at the end of World War II by Portuguese neurologist António Egas Moniz, who was even rewarded with a Nobel Prize for the procedure. He invented the technique later promoted in the US by Walter J. Freeman, of cutting through the prefrontal cortex of psychotic patients. The operation had a drastic impact on the subjects' personalities and greatly reduced their mental capacities. By 1951, as many as 20,000 such procedures had been performed in the US alone.

Lobotomy enjoyed the success it did at the time mostly because feedback was generally provided only by the families of patients. Before the procedure, relatives were confronted with a violent and unpredictable individual, who after surgery transformed into a meek and calm family member. The credit for ending these physical brain interventions can mostly be attributed to the invention of pills that achieved similar results by the use of chemicals.

Prefrontal training

To achieve successful reintegration of criminal offenders, one needs to make sure they will not repeat their crimes. The biggest problem faced by most offenders is learning how to control their impulses. They tend to be very well aware of the differences between right and wrong, yet in some situations are unable to control themselves and resist their impulses. Adequate control of every urge that crosses our mind is a key part of what enables us to function well in society. This is also what constitutes being a grown-up. Teenagers typically have more impulsive tendencies because their prefrontal cortex doesn't fully develop until their early twenties.

To prevent misguided treatments such as lobotomy, Eagleman proposes prefrontal training as a way to achieving rehabilitation. Successful prefrontal cortex training techniques function as an additional set of brain inhibitors, since the mechanisms of this specifically human brain region prevent a person taking actions that may be useful in the short term but have harmful long-term effects. These training techniques wouldn't pose any chemical or physical threat to the brain, but would in turn equip criminal offenders with the mental tools necessary for non-harmful social existence.

The biology of self-control and instinctive willpower

When survival is at stake, all our energy and attention is taken up in those few brief moments that are needed to avoid the oncoming danger. The term "stress" denotes a biological mechanism that triggers a rapid bodily response to a threat, preparing us either to fight or fly. Regardless of the actual reason for triggering the alarm, the body always reacts in the same way. In order to make us as efficient as possible in the key seconds necessary for self-preservation, all the energetically inefficient bodily functions are turned off and everything gives way either to escaping as quickly as possible or going completely numb and waiting for danger to pass.

In her book *The Willpower Instinct: How Self-Control Works, Why It Matters, and What You Can Do to Get More of It* (Avery 2011), Kelly McGonigal reports on how a group of scientists have recently described a bodily mechanism that stands in contrast to the stress reaction. It could be described as the "wait-and-think", as opposed to the "fight-or-flight" reflex. The latter reaction forces us to respond instinctively to danger and not waste precious time thinking about the situation, while wait-and-think enables us, on the contrary, to regain our self-control and think over what we are about to do. A stress reaction increases our heart rate to prepare us for action, while the wait-and-think system calms us down: instead of activating the muscles, it stimulates the brain areas responsible for self-control.

If however stress is triggered by an outside danger, the wait-and-think reaction usually appears as a result of our internal conflict between doing something that provides us with immediate relief, and following a strategy that will bring better long-term results. A nice example of this is craving a cigarette while being fully aware of how unhealthy it would be to smoke one, or watching TV when we know we should be working. The familiar sensation of internal conflict triggers a physiological mechanism that provides temporary relief and helps us reach the right decision, which is: doing what's best in the long term. In fact, the main enemy the wait-and-think mechanism protects us from is ourselves.

Respiratory sinus arrhythmia

The most reliable scientific method for observing the wait-and-think response proved to be measuring respiratory sinus arrhythmia – the variation in heart rate that occurs during a respiratory cycle. It is not widely known, yet can be detected very simply by using a mobile phone with a built-in camera to measure the heart rate and blood flow in one of your fingers.

It may sound strange, but a completely steady heartbeat is usually an indicator of stress. When the fight-or-flight response kicks in our pulse rises but its variability decreases. Moderate variability in our pulse rate can be a sign that the heart is not governed exclusively by the sympathetic nervous system, which works to prepare our body for action, but also by the parasympathetic nervous system, which calms the body down. When we breathe in, the pressure in our chest is lowered, which in turn decreases our blood pressure. This results in an automatic increase of heart rate to compensate for the unpredicted drop of blood pressure. After a few seconds, another control and regulation mechanism is triggered that slows the heart rate again.

The level of heart rate variability in relation to breathing is a very good indirect indicator of how the self-control mechanism works. If you have high heart rate variability, it means you are not stressed out and have enough strength in reserve to control yourself effectively. Experiments showed that by measuring heart rate variability one can predict which of a number of individuals would be best able to resist temptation. When they see a bottle of wine, alcoholics with increased pulse variability tend to do better at resisting the temptation than those whose heart rate variability drops when exposed to the bottle.

After the 9/11 terrorist attacks in the US, it was found that peoples' average heart rate variability dropped distinctly. A similar phenomenon was observed during the outbreak of the financial crisis in 2008. Nothing depletes the supplies of self-control faster than stress, because the physiological responses to stress and self-control are simply incompatible. The stress reflex prepares the body for immediate action by redirecting energy from the centers for long-term planning to the quick-response mechanisms.

The self-control muscle tires out quickly

For most of its evolutionary history, the prefrontal cortex has been in charge of controlling the physical movement of the human body. This part of the brain eventually increased considerably in size and started taking over additional functions. Today, it not only controls our movements, it also governs our thoughts, emotions and attention span. The prefrontal cortex is the area of the brain that makes sure we do something that might seem difficult at the moment, but will provide long-term benefits. When we pass a bakery and the smell of freshly baked treats fills the air, we usually wish we could eat some right away, but our prefrontal cortex informs us this might not be a good idea, since we have already decided to stick to a healthier diet.

It should be noted however that our reserves of self-control are limited and can run low quite quickly. We can imagine the self-control function of our brain as a muscle that can be very useful if we keep it in good shape, but can collapse entirely if we overload it– just like our leg muscles after too long a run.

Because the self-control mechanism wastes a lot of energy, our energy reserves are quickly depleted. Experiments with volunteers proved that, among other things, our ability to control ourselves depends on our blood sugar levels. If they drop, the brain automatically turns off the areas that are energetically inefficient. So it shouldn't surprise us if we become moody when we're hungry – this is a sign that we are losing self-control due to a lack of energy. That's why we should eat food that is digested slowly and releases nutrients gradually to keep a regular supply of energy running to the brain.

The author of *The Willpower Instinct* proposes an efficient method of exercising the prefrontal region of the brain. If we spend a few minutes every day concentrating exclusively on our breathing, it increases blood flow to the prefrontal region, tones the self-control "muscle" and protects us from the harmful effects of chronic stress.

The ability to control ourselves can also be improved by assuming regular responsibilities. Researchers proved with experiments that a control group of volunteers who were given an additional task (to eat with their left hand if they were right-handed, for instance) was less likely to succumb to instant gratification or immediate temptations and its members were better at controlling themselves to reach their long-term goals.

Such indirect toning of the brain's self-control "muscle" might be the reason for the extraordinary success of a social help initiative that has been developed in Latin America. This strategy for helping the poor is called Bolsa Familia in Brazil and Oportunidades in Mexico, and it could simply be described as conditional social assistance. The idea behind it is simple: poor families, typically represented by the mother, receive support in the form of cash or bank transfers if they abide by certain ongoing conditions. In Brazil, children are required to attend school and regular medical check-ups, and mothers take housework classes where they are taught more about healthy eating habits and hygienic practices.

In recent years, there has been much talk about introducing a universal basic income in Europe, which would insure that every citizen receives a monthly fixed allowance. Even though this regular help from the government might to some extent protect the weakest members of society from hardship, it doesn't seem the best strategy for systematically improving social imbalances, if the neuro-scientific findings discussed above are correct, since it disregards the importance of conditioning people's self-control "muscles" to ensure healthy development in society.

The social brain

Students tend to start forgetting what they have learned in school after an average of three months. And they are able to use an even smaller amount of this information in everyday life over the following years. In his book *Social: Why Our Brains Are Wired to Connect* (Crown, 2013), neuroscientist Matthew D. Lieberman warns that today's schools don't pay enough attention to the large system in human brains that is responsible for following and managing interpersonal relations. According to research conducted by the author, the cerebral system responsible for our social life has a powerful capacity to remember, but is activated in slightly different circumstances to those we are used to in schools.

When researchers told a test group to remember some information about random people on which they would later be tested, the group did worse than another which was told to get a general impression about a person based on the information given. Even though the people in the second group weren't told their knowledge would be tested, they did a better job because they used the area of the brain responsible for social contact.

It also turned out that we have less difficulties learning new things if we know we would need to pass on our newly acquired knowledge to somebody else and not only use it in a test. When we think about presenting the content of our studies in an imaginary dialogue, we use a different mechanism in our brain and are able to acquire new knowledge easily and more permanently.

More detailed research into these mechanisms of remembering and learning is continuing, and we can soon expect the learning process used in schools to change radically as a result. There will probably be fewer long lectures where students only listen to teachers and take notes, with the expectation that they remember most of what has been said and then reproduce it in an exam. According to the author of *Social,* schools will soon use more practical techniques that involve interpersonal social contacts, because they include more capable "social" parts of the brain in the learning process.

People are social creatures

People are capable of much more complex forms of interpersonal collaboration than any other biological species. We have developed a complicated social system along with a method of role distribution that surpasses anything else found in nature. For this we need to thank specific mechanisms in our brains that put a great amount of effort into following and managing our interpersonal relations.

In past centuries people were generally convinced they were guided by animal instincts, yet believed that individuals might control their "animalistic nature" with reason and function as free beings, each responsible for his or her actions. A human was supposedly a combination of a high-performance computer and a creature striving for much comfort and little pain. But this oversimplification doesn't take into account that many activities in the human brain take place somewhere in between the poles of conscious analytical reasoning and basic physiological impulses.

An important reason for the enlargement of the brain in evolution has been its adaptation to ever more complex social relationships. Animals that interact more with each other are safer from predators and other dangers in the wild, but are at the same time subjected to conflict within their particular group. All the individuals in the group can never be equals and some will always need to work to keep their status.

In order to optimize the way we function in groups, our brains have highly developed mechanisms that enable us to foresee other people's actions and reactions. One important mechanism allows us to make intuitive guesses about other people's intentions based on their behavior, facial expressions and other external signs they project. Another crucial function allows us to see the world through the eyes of others - which helps us understand them better and predict their actions.

Scientists usually study which areas of the brain are triggered by a certain activity, but in 1997 Gordon Shulman looked at the opposite side of the question, and asked which areas of the brain are stimulated when an activity stops. He discovered that certain regions of the brain are activated when we stop doing something or when we do nothing in particular. The default network, as these regions are known collectively, is activated at every opportunity or when our brain activity isn't focused on any other specific task.

Matthew D. Lieberman and his colleagues recently observed that this default brain network, which hasn't yet been properly researched, seems much like the area of the brain that is switched on during social tasks. Lieberman therefore concludes that people are social beings by nature precisely because of the connection between our default brain activity and the thinking which manages our personal relations. We need to be aware that such social thinking is not analytical reasoning about our interpersonal relations, but mental activity which occurs spontaneously on an intuitive level.

Altruism

The principle of reciprocity is one of the most powerful social norms. If someone does us a favor, we feel obliged to return it. This is why smart salesmen who sense that we might spend a lot of money on their products offer us cups of coffee and other small gifts, for which we intuitively feel obliged to pay in some way. And if we do buy that car in the end, the seller's tiny investments are suddenly worth a lot more than their intrinsic value.

Interpersonal collaboration is rooted so deeply in our nature and the structure of our brain that we tend to opt for it even if we don't benefit from it directly. Research shows that we don't collaborate only because it's good for us in the long term, but because collaboration is a source of comfort and wellbeing in itself.

Lieberman defines altruism as an activity that doesn't have any long-term material benefits or can even have negative consequences for those who put other people's needs before their own, but which people still do anyway. He draws an analogy with sexuality, where sex is an activity we normally don't pursue in order to achieve conception but because we enjoy it for what it is. The same, he says, goes for altruism. We usually don't do good deeds in order to be repaid but because we enjoy doing something good for someone else.

Lieberman adds, however, that people are by nature both selfless and selfish. Mammals' brains have a built-in need to care for others who are not necessarily their closest relatives. Selfless acts bring us satisfaction regardless of the reward.

Social pain

Knowing that we matter to other people, that they care about us or look up to us is essential to our wellbeing. Many studies have proven that hearing a word of praise about ourselves triggers a similar response to eating a dessert. The brain region connected to the pleasure of eating chocolate responds in the same way when we realize that we're appreciated.

Yet the same brain mechanisms aren't shared only by the physical and social sources of pleasure, but also by physical and social sources of pain. We intuitively feel that social pain must be very different to and less real than its physical counterpart, but as far as the brain is concerned the mechanisms are very similar and partly the same. Our brain feels a threat to our status in society as a social pain. By triggering the same mechanisms that are active during our experience of physical pain, our social pain warns us that good relationships with other people are essential to our survival. Feeling that we are being treated equally and fairly is especially important.

Some children with inborn deformities, who can't feel physical pain, rarely live longer than a few years. They aren't able to feel when they get hurt and sooner or later succumb to their injuries. The feeling of pain is therefore essential for our survival. Things are not much different with the pain caused by social life. The only difference is that individuals who can't feel this "social pain" are still able to live a long happy life. But if most people didn't care for the opinions of others, our society wouldn't last long before falling apart.

The brain: detective on the trail of a serial killer

More than half a century after his death, the German-Estonian biologist Jakob van Uexküll is probably better known among philosophers than natural scientists. He was interested in how different living beings subjectively experience themselves and their surroundings. To describe the specific experience of each animal, he introduced a German word, "umwelt", that became an international term and usually isn't translated.

In Uexküll's studies of individual beings' different subjective experiences, the main role of test subject was surprisingly given to ticks. According to the knowledge available at the time about the physiology of this fairly dangerous animal, ticks live in an exceedingly simple version of the world. Today we know that world is a slightly more complex matter even for ticks but this is not very important for understanding the umwelt principle.

The world according to a tick

Uexküll described the tick as an example of a very simple subjective image of the world because it uses only three signals to define its surroundings and orient itself accordingly. It has no eyes, but its photosensitive skin leads it to move towards the light. This enables the tick to climb as far up as it needs to from the ground, for instance to the top of a grass blade, where it will wait for a host to suck blood from and feed upon.

The tick is deaf and blind, relying on its sense of smell to spot its victims. Mammals' skin glands give off a specific smell, including butyric acid, a chemical compound that activates the tick's limited senses. "The smell of mammals", or rather, the chemical compound producing this smell, is released with sweat and is taken by the tick as a signal to let go of the grass, hopefully landing on the animal as it passes. If it succeeds, all that remains for the tick to do is to follow its sense of heat to a point on its host's body where it can have easy access to blood.

According to Uexküll, every being has its own specific umwelt, arising from its actual sensations and other biological traits. There is no single subjective world for all creatures but a multitude of different worlds. People are nevertheless convinced that our image of the world, probably the most complex on earth, is universal and necessary, but Uexküll's approach proves that this common presumption is wrong.

How a certain object is perceived depends greatly on the particular subject or perceiver. Every animal possesses a limited ability to distinguish the stimuli in its surroundings, which forms the basis for its image of the world. The same goes for people.

Only animals with eyes distinguish between the visual and tactile space around them, while some animals don't have the ability to tell these two sensations apart. In visual space, the size of the object changes with distance, while to the touch it remains the same. This distinction is only available to creatures with eyes.

Uexküll's understanding of time is also very interesting. He defines the basic unit of time for each being as the longest interval of zero change in its subjective world. For a human that's about one eighteenth of a second, the shortest interval our senses can register. Thanks to this we can see a homogenous visual flow on the cinema screen, instead of thousands of pictures projected one after the other.

According to Uexküll, a tick's basic time interval will be considerably longer because its senses take longer to perceive change. If there is no mammal smell nearby, a tick's subjective representation of the world remains static, meaning that time doesn't flow for it. The time interval lasts until the tick receives a signal from one of the other two channels available. Uexküll acknowledges this concept of time might oversimplify, yet he continues to use it in order to present his theory clearly.

The intimate world of ticks serves as a good introduction to the human perception of the world. In his book *Vision and Brain: How We Perceive the World* (MIT Press, 2012), James V. Stone warns that the sense we have of continuously perceiving our environment is actually an illusion created by our brains. We don't watch constantly; rather, we just have a constant sense that we are watching.

If we would actually see what our eyes perceive at a given moment, our image of the world would turn to chaos. According to Stone, we walk around functionally blind most of the time without being aware of it. We see by momentarily focusing on details around us, while screening our environment incessantly. At any particular moment we can only see clearly a small part in the middle of our field of vision. We are oblivious to this because the brain supplements the information missing from the picture in front of us with data we already possess about our surroundings.

The human retina contains sensory cells that transform light into electric signals and consists of as many as 126 million of these units. Each unit senses incoming light differently, forwarding information about it to the brain. The optic nerve that transmits signals from the eyes to the brain enables only a million connections, which means that less than 1% of all the information gathered by the retina ever reaches the brain.

The origin of eyes in the evolution of life

Nevertheless, the eyes transmit enough information for the brain to construct an appropriate image of our surroundings. Much information gathered by the retina turns out to be duplicated or redundant. The process is similar to the way a picture file from a digital camera may be shrunk without noticeably reducing the quality of the photo.

The image of the outside world reflected by our retinas is full of mistakes and confusion and still lacks a lot of information we would need to make sense of it. As a result, the brain must fill in the resulting blanks to stop us feeling as if we were surrounded by a giant puzzle with key pieces missing. If the brain's visual processing system reflected what our eyes really see, our sight would be very foggy, with many grey areas. Because our surroundings don't usually go through dramatic changes on a daily basis, our brain can rely on memory and previous experience to fill the gaps in our visual comprehension of the world.

While we watch something, our brains work hard to establish which objects should trigger which reactions in our bodies' sensory cells. James V. Stone says that our brain is a sort of detective, trying to find out who killed the victim – the 'unedited' image of the outside world received by the retina. The brain is constantly checking different possible views to trigger the right sensory reactions to the stimuli it interprets. What we see in the end is the result of eliminating some of the hypotheses.

Even though life on Earth is 3.8 billion years old, living creatures managed to survive most of it without eyes. The first simple set of eyes appeared only during the 'Cambrian' explosion in evolutionary history about 540 million years ago, when many different life forms emerged.

Eyes reached their peak of development in the context of the natural fight for survival between predators and their prey. While prey tried coming up with more effective ways of hiding from their natural enemies, hunters needed to develop ever more efficient eyesight to see through their targets' increasingly sophisticated camouflage.

This led to the evolution of spatial vision, and other highly effective methods of sensing every detail in the environment, down to the tiniest movement. Because many different eye-shapes are observed in the animal world, scientists believe that eyes developed independently more than once in the course of evolutionary history.

NUMBERS

The genius who never existed

Nicolas Bourbaki is without doubt one of the most influential mathematicians of the twentieth century. His books laid the foundations for the entire field of modern mathematics and profoundly influenced future generations of young mathematicians, as well as the way the subject was taught in schools and universities. He is accredited with the famous "new mathematics" movement that introduced children all over the world to numbers and equations with the use of set theory. The influence of his technical publications, which amounted to no less than ten thousand pages, extended beyond professional circles and into the humanities and social sciences through the structuralist movement.

But the story wouldn't be so special if Bourbaki was in fact a real person. It is hard to say that Nicolas Bourbaki never existed, but his life was nothing like that of other famous scholars. Bourbaki was merely a pseudonym, used collectively by a group of young French mathematicians who published university textbooks in the 1930s.

World War I left a great scar on French academia. According to some sources, almost half of the university students who graduated after 1910 were wiped out in the conflict. In the following years, professorial chairs were therefore taken up by young graduates who had finished their studies during or only just after the war. This generation made big steps in the development of a number of scientific fields. It was the time when young physicists started developing quantum theory, Albert Einstein having provided a completely new model of time and space just a few years earlier: all in all this period can easily be described as one of the most creative and prolific in the history of science.

A group of young French math professors, most of whom were under 30, initially got together to standardize the basic university math courses they were giving at various universities in France. They agreed that it was high time math education was drastically reformed, due to outdated teaching programs and textbooks that required teachers to make individual updates, all of which led to a confusing array of different pedagogical approaches.

Under the chairmanship of André Weil, professor at Strasbourg University and brother of the French philosopher Simone Weil, the group's first meeting took place at noon on December 10, 1934 in a café in Paris' Quartier Latin. The group was initially composed of six members, later amounting to an average of twenty, and promptly set out on an ambitious mission. They planned to co-write a university textbook on mathematical analysis, a branch dealing with theories of integration, derivation and differentiation, all of which are basic mathematical tools used in engineering.

They decided not to edit the existing textbooks but to make a fresh start by rewriting them completely, laying new foundations for a new era in math education. In any other circumstances, the group would probably have had to immunize itself to sneering remarks from older colleagues, setting forth a bewildering array of reasons why this sort of reform was impossible. But this group of aspiring scholars was in no way dependent on the older generation and they were able to put all their strength into the incredibly ambitious plan.

We can claim without exaggeration that these young professors took on a task that could be compared to the work done by Euclid two thousand years ago, when he used the axiomatic method to rearrange the entire stock of ancient mathematical learning. The group began by setting some fundamental rules that guided their work for decades to come. They agreed to make decisions unanimously and to approach mathematics from the most abstract of possible angles, relying mostly on set theory.

The idea of creating a pseudonym, Nicolas Bourbaki, was born during their first week-long congress, in the summer of 1935, in the French countryside. The name alluded to a prank played by a young student mathematician a few years before. He entered the lecture hall dressed in a general's uniform and wearing a fake beard, and started to demonstrate something he called "Bourbaki's theorem" to a class full of astounded freshmen. To break up long discussions of abstract mathematics, the congress participants prepared a solemn christening ceremony for Nicolas and wrote a report about it for the newspapers. A few years later they even printed out marriage invitations for the wedding of his daughter Betti and eventually drew up his entire family tree, making him seem more real still.

Soon after the first works by Bourbaki works were published, the group of French mathematicians was awarded a mention in the Encyclopaedia Britannica. The author of the entry explained Bourbaki was a collective pseudonym and soon received a postcard from an ashram in the Himalayas beginning: "You miserable worm, how dare you say that I do not exist?" signed: Nicolas Bourbaki.

The man who proved that not everything could be proven

Kurt Gödel is considered one of the most important logicians of all times. His proof of unprovability in mathematics ranks among the greatest scientific achievements of the twentieth century. Before his ingenious findings, many mathematicians were convinced that the proof of all mathematical truths lay in a careful selection of certain fundamental rules called axioms. Mathematics was understood as something like a computer program, capable of deriving every theorem from a careful selection of basic definitions. But Gödel showed that the human mind would never be able to create such a perfect mathematical system.

Kurt Gödel, nicknamed Herr Warum (Mister Why) by his family as a child, due to his incredible thirst for knowledge, was born in 1906 in Moravia (today's Brno in the Czech Republic), where his father owned a textile factory. He began to study physics at Vienna University and later changed to mathematics, when in 1927 he met Adele Nimbursky, a bar dancer and a few years his senior.

His family opposed the liaison but Kurt didn't succumb to their pressure. He insisted on his choice, but was still forced to wait more than ten years before they could finally marry. Even though marriage between an uneducated bar dancer and a mathematical genius seemed destined for failure, it turned out that the curious logician and the unschooled yet intelligent dancer were a great match. Adele and Kurt lived together until his death at the end of the 1970s.

At one point after the rise of Nazism, Kurt was attacked by a group of Nazi hooligans in Vienna, who probably thought he was Jewish. He was saved by a cool-headed lady who stood up to the assaulters and defended the fragile scientist, who preferred saving his energy for abstract mathematical derivations than using it to show off his muscles.

When he was summoned for a medical check at the beginning of World War II, before being drafted into the German army, he decided to escape. He and Adele ran away and made a journey taking several months across Russia and Japan, all the way to North America where a number of Kurt's fellow scientists also sought refuge from Nazi Europe. The couple left Austria on January 18, 1940 and arrived in San Francisco on March 4 the same year.

In the USA Gödel took up a position at Princeton's *Institute for Advanced Study,* nicknamed a Scientist's Retirement Home by Adele for the sanctuary it offered many elderly European refugees. Gödel had visited the Institute on several occasions as visiting professor in the 1930s and was familiar with the center's working environment.

Albert Einstein was without doubt the most famous researcher at the Institute, and he and Gödel soon became close friends. They walked together from their homes to the Institute and back every day, sharing their thoughts. Young employees later remembered Gödel as the only one who conversed with Einstein as an equal; nobody else was able to interact so casually with the world famous physicist.

Gödel filed a request for American citizenship during World War II, which required him to pass a number of exams and interviews, with functionaries checking each candidate's knowledge of the American legal system. As a dedicated logician, he meticulously prepared for each interview, starting with a detailed analysis of the American Constitution. Doing so, he came across a logical loop that supposedly left open the possibility of legal dictatorship in the USA. He initially planned to present his findings to the functionaries at the naturalization interview but his witnesses at the procedure, Albert Einstein and the economist Oskar Morgenstern, wisely stepped in and kept him out of trouble.

Gödel had always been known for approaching problems from a slightly different angle to most of his peers. When he was asked on a certain occasion to contribute an article for a collection of scientific papers on Einstein's work, he provided an original solution to Einstein's equations from the general theory of relativity that discusses space and the possibility of travelling back in time.

At the age of seventy, after almost forty years of living in the USA, Gödel finally retired. A year after his retirement, Adele underwent surgery, followed by a long hospitalization. For Kurt, who was not used to being alone, this temporary separation from his wife had disastrous effects. He had been suffering from various mental troubles as it was, but at that point he stopped eating all together and was convinced that somebody was trying to poison him. When his state deteriorated, he was taken to hospital where he continued to refuse food. After two weeks, on January 14, 1978, he died of malnourishment. Just before he died, he weighed no more than 30 kilograms.

The Incompleteness Theorem

Kurt Gödel is famous for many important new findings in mathematics and logic, but his greatest and most influencing accomplishment is without a doubt the proof that even in mathematics not everything can be proven.

In 1931 he published a ground-breaking paper proving that in any mathematical system enabling at least basic arithmetic operations, we can construct a formula that claims to be unprovable and can't be solved within the given set of axioms. The existence of such statements poses a great problem because it leads to contradictions. If such a statement is proven, it means it is false since it claims it can't be proven, and yet we proved it, creating an obvious crux.

We can illustrate the meaning of Gödel's finding with the ancient allegory of the famous liar paradox. If I say: "I am lying", my statement can neither be true nor false. If it was true and I really was lying, then the conclusion would be that the true statement was: "I am not lying", which doesn't match my actual statement of "I am lying". The same complication arises with Gödel's formula which, unlike the liar paradox, is applicable to all mathematical systems.

Gödel's incompleteness theorem proved that mathematics can't be set up as an integral logical system, since there will always be a statement that can't be proven true or false exclusively by the means of formal derivation. This is why mathematics can never be fully encompassed by any complete system of axioms.

What does that mean?

Soon after Gödel's paper was published, the meaning of its abstract findings was subjected to further interrogation. Among those questioning it was British mathematical genius Alan Turing. While trying to find a more illustrative and tangible way to present Gödel's abstractions, he came across his idea for the Turing machine – a theoretical predecessor of today's computer.

If we look at the computer as a logical machine that abides by any set of pre-conceived rules, we can compare the incompleteness theorem to a realization that there will always be a problem that a computer won't be able to solve. Turing proved that an analogue problem, as shown by Gödel's incompleteness theorem, in the context of computer science turns out to be a problem of the finite number of steps which lead to an algorithm or a program. A program can either end with a solution or it can go on forever. On the basis of Gödel's derivations, Turing additionally proved that the algorithm itself does not tell us whether the program will perform its tasks infinitely or eventually reach a final solution.

Several further questions had arisen as to the possibility of applying Gödel's findings in mathematics and logic to understanding the processes of human cognition. Our mind functions in accordance with the principles of logic. It can actually be perceived as a sort of "computer" and therefore abides by the principles of Gödel's findings.

Yet in everyday life we don't encounter many difficulties with the paradoxes predicted by Gödel for formal systems. What does that mean? Is it possible to conclude that human beings are not machines and that human cognition can't be translated into a formal system of logic? Is there more to the mind than reason? These are questions that Gödel himself posed in the last decades of his life but without any conclusive answers.

Gödel's notes containing the arguments he developed in logic and mathematics are very clear and concise, as opposed to the notes on the implications his incompleteness theorem might have for human cognition. These are copied, crossed out, colored and drawn on, illustrating Gödel's struggle with this seemingly unsolvable issue.

The mysterious disappearance of Ettore Majorana

The Italian genius of physics, Ettore Majorana, boarded a ship to Naples in Palermo on March 26 1938 and disappeared without trace. At the young age of thirty-one, he was already considered a behemoth of theoretical physics. His mentor Enrico Fermi even compared him to Newton. The young physicist's body was never retrieved. Despite the possibility he committed suicide, it was extremely unusual for a suicide victim to board with a passport and a sum of money worth tens of thousands of euros in today's currency.

The mystery surrounding Majorana's death deepened when a number of his acquaintances started to claim they occasionally saw him from afar on the street, even after he had been given up for dead. There was talk of him entering a monastery, which led his family to ask for him in Christian residences and refuges, carrying his picture with them, but without success. Others believed that he travelled all the way to Argentina, where he was seen several times and was reportedly even caught in a photograph. As is often the case with unusual scientists, theories went around that he had been kidnapped by the Russians and forced to build an atomic bomb, or in some cases even by Martians. The most amusing of the theories however, was the science-fiction hypothesis that he had invented a way of entering a parallel universe.

The Via Panisperna boys

The first encounter between Majorana and Fermi, when young Ettore first entered the famous Physics Institute on Via Panisperna in Rome, is still a source of inspiration for many young scientists today. It was there that Fermi, like Niels Bohr in Copenhagen, systematically brought together young geniuses in mathematics and physics, providing them with a peaceful environment that enabled them to tackle in depth questions they considered crucial to the future of science. They were able to have a lot of fun, strange as that might seem to someone who doesn't spend day after day with equations. They raced each other at solving an overwhelmingly complicated differential equation or a problem in atomic physics. It was the time when Italy, largely because of Fermi's group, was at the summit of international science.

When Ettore first visited Fermi at the Institute, the latter was in the middle of several months of calculations to solve an important problem in atomic physics, work which he presented to his young colleague. It was in the pre-computer era and complicated equations took enormous amounts of time and energy. Fermi explained to Majorana why for the time being he was only able to solve a smaller part of the problem. Ettore listened carefully and asked a couple of questions. Then he vanished. The next day he came back to the Institute and asked Fermi to show him again the results it had taken several months to obtain. He took a scribbled piece of paper out of his pocket, did some additional calculations and praised Fermi for having obtained the correct results. He walked over to the blackboard and, before an audience of astounded physicists, wrote down a mathematical transformation that changed Fermi's complicated problem into a universally known equation, now found in textbooks across the world.

The pioneer of many groundbreaking ideas he never published

Majorana was born in August 1906 in Catania, Sicily. It soon became obvious that he was an exceptional child. He was remarkably bright and able to solve complicated equations from memory. At the same time he was very timid. His mother came from a wealthy family and his father established the city's first telephone company, later nationalized during fascist rule. The family moved from Catania to Rome where Ettore's father worked as a telecommunications inspector.

Ettore began his studies at the early age of seventeen to become an engineer but changed to physics after taking his degree and soon earned a PhD. His first scientific paper was published in 1928, followed by a handful of other articles mostly in Italian. He was extremely introverted, which presented a major personal problem and most probably contributed to his reluctance to communicate new and important findings to the general public. He was reportedly the first to hypothesize the existence of the neutron but despite his mentor's encouragement, never published his work. The Nobel Prize for the discovery of the neutron was awarded to the English scientist James Chadwick in 1935. Perhaps Majorana thought it was too simple and obvious to report it to other scientists.

In 1933 Majorana was awarded a scholarship and travelled to Leipzig to meet Heisenberg- There he worked with other key physicists of the time. He visited Bohr in Copenhagen and published a few papers on atomic physics. Today, his theoretical work on the mass of neutrinos has proven to be of extreme importance since it has been confirmed only recently.

Fermi once wrote about three types of scientists. Most of them work as hard as they can and don't get all that far. A small number of scientists contribute to truly great discoveries that form a pillar in the progress of science. According to Fermi, this group includes the likes of Heisenberg. And then there are geniuses such as Galileo and Newton. According to Fermi, Ettore Majorana belonged in this rarest category of talent, but unfortunately lacked the common sense to publish his brilliant ideas.

The fateful journey to Palermo

On Friday March 25 1938, Majorana wrote two letters in his room in Naples. The first letter was addressed to his superior, the second to his family. He had recently become a full professor at the University of Naples where he was supposed to hold a lecture on quantum mechanics the next morning. Sadly, he never made it to the lecture. At 10.30 p.m. he boarded a ship to Palermo which was to arrive the next morning.

On Saturday, his superior Antonio Carrelli received a telegram saying: "Do not worry. Letter on the way. Majorana." The letter soon came with the mail. In the letter Ettore excuses himself for his sudden disappearance and asks Carrelli to remember him with kindness. Mayorana left the second letter on the table in his room. In that letter he asked his family not to mourn him for more than three days, sought their forgiveness and requested that they keep him in their thoughts. Even though both letters could be interpreted as suicide notes, those who read them said they were written in a confident manner, quite unlike the majority of suicide letters. In neither letter does Ettore mention death, only disappearance.

Ettore surprisingly got off the ship in Palermo the next morning and booked a room in a hotel. He wrote another letter to Carrelli on hotel writing paper, saying he hoped Carrelli had received his previous letter and telegram. He said that he was coming back home the next day. He had decided to refuse the teaching position at Naples. Ettore ends his letter by saying he is no hero from an Ibsen play, adding that his case is different and offering to give an explanation to Carrelli upon his return. On Saturday evening his ship left Palermo for Naples. Ettore bought a ticket but never arrived. Ever since then he has been listed as a missing person.

On Wednesday, March 30 Carrelli wrote a letter to the chancellor of the University of Naples, describing the incident. He quoted the unusual letters and the telegram he had received and reconstructed Majorana's moves up to the Saturday evening when he disappeared without trace. He mentioned notifying his family in Rome and that his brother was coming to Naples on Tuesday to organise a search. But even after a police investigation, they weren't able to prove Ettore was still alive.

It seems that the mysterious case of the missing genius physicist is still newsworthy. In March 2011 the Italian media reported that the Majorana case was being reopened in Rome. The prosecutors had supposedly received some reliable photographs and a witness reportedly met Majorana in Buenos Aires soon after World War II. On June 7, 2011 *Corriere della Sera* reported that the *Carabinieri* had carefully studied a supposed photograph of Majorana dated 1955 and confirmed that the man in the picture greatly resembles the physicist who had gone missing in 1938. He was reportedly living in Argentina under the name of "Mister Bini".

Scientists on vacation

It is said that the world's top scientists are always working, even while on vacation; although not of course in the same way as CEOs who must be available 24/7, taking business calls on the beach while wading into the sea, their thoughts still trapped within their office walls. Scientists, though, keep working even when science is the last thing on their minds.

We can learn from the history of science that some of the most innovative ideas were discovered just as their discoverers relaxed from painstaking work. Let us consider one such event that took place a little less than a hundred years ago and caused a major revolution in physics.

Summer island vacation

Helgoland is a small island in the North Sea, not far from Hamburg. It boasts distinctive steep red cliffs and lush green grassland, offering a perfect escape from the summer heat. Seeking refuge from his annual attack of hay fever, the young German physicist Werner Heisenberg arrived at this idyllic piece of land in June-1925. A few days later, dipping his feet in the cold seawater, he was struck by an idea that brought an important new turn in the development of quantum physics. To understand fully the importance of young Heisenberg's idea, we need to shed some light on what was going on in science at that time. But first, a short introduction to the hero of this summertime story is in order.

Werner Heisenberg was a typical representative of the generation of young physicists who concluded their studies after the First World War. Their characteristic traits included a certain lack of respect for senior colleagues. This young generation found it essential to figure out new approaches to scientific problems, and regarded the old academic establishment largely as an obstacle.

Mens sana in corpore sano

Heisenberg's father was a classical philologist. He introduced his son to ancient philosophy and literature at an early age. Young Werner joined the German Youth Movement, whose members lived by the credo "Mens sana in corpore sano (healthy mind in a healthy body)". Despite his love for classical music (he was an exceptional pianist), he became increasingly interested in the natural sciences.

Heisenberg studied in Munich with professor Arnold Sommerfeld, who invited him as an up-and-coming scientist to attend a physics conference in Göttingen in 1921. There the most prominent scientists discussed all the burning issues of that time. Heisenberg initially planned to be a mathematician, but changed his mind after a long conversation in Göttingen with the Danish physicist Niels Bohr. He was so fascinated with the new problems in atomic theory that he decided to pursue a career in physics. He realized that abstract mathematics was becoming increasingly important for the understanding of new problems in physics, and the connection between purely abstract mathematical thought and the real world intrigued him the most. When he obtained his PhD in 1924 with professor Sommerfeld, he went to Denmark to work with Bohr on problems raised in the newly established field of atomic physics.

When a brewery subsidizes alkaline science

Niels Bohr's long-time wish was to create a working environment where scientists would be free to discuss technical problems regardless of the hierarchy between professors and students. In 1920 funds from Danish companies (including the Carlsberg brewery) helped make his wish a reality and the Niels Bohr Institute in Copenhagen was founded. Bohr gathered brilliant young students from Europe, America and Russia to work together in the new field of atomic physics.

Heisenberg fitted perfectly into this environment. He was surrounded by geniuses who in a few years' time became leading experts in their fields. What they all had in common was their brilliant knowledge and intelligence accompanied by an enormous amount of confidence, nurturing a strong belief that they were always right. They weren't discouraged by society's wider lack of interest in their work. Nothing could undermine their conviction that they were headed for new discoveries that would completely transform physics and our understanding of the world.

The crucial summer idea that changed physics forever

After a year at Bohr's institute, Heisenberg took a position as assistant with Max Born, who then chaired the Göttingen University Institute for Theoretical Physics. But only a few months into his new job, he decided to take a short break to order his thoughts. Upon arriving on the tiny island in the north of Germany, an idea occurred to him that set a new foundation for the theory of quantum physics.

Like the majority of physicists at that time, Heisenberg directed most of his energy to explaining the way atoms emit light. The central problem then was how to explain the fact that atoms only diffuse light in certain colors. What goes on inside the atoms to make them produce not the full spectrum of light but only certain colors?

How abstract mathematics explained atomic behavior

Heisenberg's illuminating insight, which flashed into his mind during his vacation on Helgoland, revealed that the key question might not lie with what was going on inside the atom, but rather with how to formulate mathematically what was already known about the atom. What if we stopped searching for mechanisms to explain the functioning of atoms, and instead used numbers to describe the colours they emit? He quickly found out that these descriptive numeric sequences formed clear formal mathematical patterns of connection, which meant they could be calculated and predicted.

Upon returning from his vacation, Heisenberg shared his findings with Born, who immediately noticed that the numeric sequences in fact form mathematical matrices. As is well known, a matrix is a generalized idea of a number, an idea represented by, for example, a square list of numbers, consisting of the same number of rows and columns. Mathematicians have long known the rules for the addition and multiplication of matrices. To Born and Heisenberg it seemed that matrices, and not numerals, offered the right language for describing atomic behaviour.

Then Niels Bohr sets off on vacation

Heisenberg's groundbreaking idea triggered a wave of debates and theories on possible improvements. Even though physicists were swift to work out equations on how to predict the color of light absorbed or emitted by individual atoms in different circumstances, they weren't able to understand fully the processes this phenomenon involved. And they became increasingly frustrated.

The scientists at Bohr's Institute tried hard to explain these new findings in the atomic world. Bohr, Heisenberg and other physicists spent a year at the Copenhagen Institute, discussing how even to begin interpreting quantum physics. They were on the verge of giving up, but Bohr's optimism kept them going. Yet by February 1927 Bohr himself was so exhausted that he needed a break. Since it was the middle of winter, he decided to go on an alpine touring trip to the Norwegian mountains north of Oslo.

The origins of today's textbook equations

While on holiday, alone with his thoughts, the basic interpretation of quantum physics, the very same that still dominates today's physics textbooks, occurred to Bohr. The basic premise of his realization goes as follows: each time we ask a question about nature, we need to specify how to interpret its answer. As Bohr said himself: "It is wrong to think that the task of physics is to find out how nature *is*. Physics concerns what we can *say* about nature."

While Bohr was away, Heisenberg too made a revelatory finding, following a long discussion with his colleagues, which contributed further to the foundations for interpreting quantum physics. Both he and Bohr reached very similar conclusions. Heisenberg came up with the famous uncertainty principle, while Bohr conceived the principle of complementarity. Both principles form the core of the Copenhagen interpretation of quantum mechanics.

Please note that leaving for vacation doesn't necessarily mean you will experience a profound scientific insight. It is what you do between your vacations that really matters.

The birth of electromagnetism

It doesn't happen very often that one stumbles upon a new scientific discovery while lecturing in an auditorium filled with students. It is even more unusual for this new discovery to be the outcome of a simple experiment set up especially for the lecture, but also conducted for the very first time during that lecture. On April 21, 1820, Danish physicist Hans Christian Oersted entered the history of science by conducting a very simple experiment in front of his students, and releasing an avalanche of new scientific discoveries that now form the basis of much modern technology.

Right before the eyes of a sleepy, listless crowd, Oersted was the first man in history to show that electricity and magnetism are interconnected natural occurrences. Oersted noticed that an electrical current transferred through wire causes a magnetic compass needle to twitch away from the north, thus proving a direct relationship between electricity and magnetism. As he later wrote, his groundbreaking discovery "made no strong impression on the audience."

A physicist holding a PhD in Kantian philosophy

Hans Christian Oersted was born in 1777 in the small Danish town of Rudjobing. His father, Soeren Christian Oersted, was a pharmacist, and young Hans Christian soon became fond of science and experiments while helping at the family store. Together with his brother, he received his initial education at home and continued his studies with considerable success at Copenhagen University. It was there that he obtained his doctorate in 1799, surprisingly not in natural sciences but in Kantian philosophy, with a dissertation entitled *The Architectonicks of Natural Metaphysics*. His strong interest in philosophy had an important influence on his later scientific work, including the discovery of electromagnetism.

After gaining his PhD, Oersted received a scholarship that enabled him to travel around Europe and learn about the latest achievements in science and philosophy. On spending time in Germany, he became enthused by the movement surrounding the German philosopher Friedrich Wilhelm Joseph Schelling. The Romantic Movement strived to surpass a mere mechanical way of understanding nature. Throughout the Enlightenment, it was highly popular to perceive nature in terms of a mechanical model, which proved insufficient since it couldn't describe the incredible diversity of natural occurrences.

Supporters of the *Naturphilosophie* movement interpreted nature as an organic whole, maintaining that essential elements are lost when it is treated partially. According to their beliefs, nature couldn't be defined as the sum of its constituent parts. Today, this movement is regarded as a kind of romantic science, but one cannot deny its importance for later scientific developments in the nineteenth century, including the area of electricity and magnetism.

An electrical current causes a magnetic needle to twitch

Oersted's research was greatly influenced by the Naturphilosophie movement, which claimed that the forces seen at work in nature were in fact different manifestations of one essential power. This idea has much in common with today's effort to unify all the known forces of nature in a single theory of physics, a quest dating back to Einstein's time. Put briefly, one of the central questions in physics today is how to unify the theories describing micro- events, known collectively as Quantum Physics, with those explaining events taking place on a massive scale in space and time, which rest in the domain of the General Theory of Relativity. Similarly, nineteenth-century physicists wondered whether electrical and magnetic phenomena might be manifestations of a single "electromagnetic force".

Soon after returning from his journeys across Europe, Oerstad became a professor of physics at his home University of Copenhagen. He reportedly possessed an immense talent for teaching, which resulted in an increased amount of pedagogical obligations and an average lecture span of five hours a day. Each month he organized a special lecture to introduce new scientific findings in areas he followed. During one of these lectures in April 1820, the one where he made his famous discovery, he spoke about electricity and magnetism, and prepared a simple experiment to test for any possible magnetic properties in electricity.

He connected a simple battery, invented two decades earlier by Alessandro Volta, to a wire that ran above a magnetic needle. When released, the electric current ran through the wire and caused a tiny flicker of the needle. The change was barely noticeable and it was unclear whether it was caused by the electric current or some other interference. Only when a stronger current was released, the deflection became obvious enough to confirm a connection between electricity and magnetism. Oersted later recalled having considered aborting the experiment at the time, since he hadn't been able to test it beforehand, but nonetheless decided to proceed. The students therefore had a once-in-a-lifetime opportunity to witness a historic event in science, but judging from their lack of enthusiasm, Oersted wrote, those in the back rows must have slept through it.

Interestingly enough, Oersted didn't publish a report of the experiment immediately. Much still remained unclear and his theory about what really occurred during the experiment was in fact never confirmed. But nevertheless, the fact that the electrical current disturbed the needle remained extremely significant. No sooner than the summer of that year, Oersted wrote a paper describing the event, entitled "Experiments on the effect of a current of electricity on a magnetic needle". Though originally written in Latin, it was soon translated into a number of European languages to make it accessible to a wider circle of readers. The article begins as follows:

> The first experiments regarding the subject which I intend to explain here were made by me last winter, while lecturing on electricity, galvanism, and magnetism, in the University.

Praise rained down on the author from across the world and anyone who knew anything about science and technology saw a new era approaching. Nobody knew exactly what the discovery of electromagnetism actually meant for humanity, but from today's perspective, in a civilization dependent on equipment exploiting the relationship between electricity and magnetism, the historical importance of this lecture in Denmark can't be overstated.

The beginning of the era of electromagnetism

A few years after his groundbreaking paper was published, Oersted set out on another journey across Europe, meeting many important physicists of the time, among them Michael Faraday in England. Upon his return he founded the College of Advanced Technology in Copenhagen, a predecessor of today's Technical University of Denmark.

Oersted's discoveries weren't only significant in terms of the connection between electricity and magnetism. One of his key findings was the existence of a new type of force that wasn't an ordinary force of attraction, emanating from a known source and moving in all directions, such as gravity or electrostatics. The magnetic force affecting the compass needle through the electrified wire was completely different from any physicists knew of and could describe mathematically.

A few months later, the French genius of physics André-Marie Ampère transformed Oersted's discovery into a corresponding mathematical form. Inspired by the pioneering experiments of the Danish physicist, he spent the following years studying electromagnetism in great detail and in 1827 published an extensive study dubbed *The Principia of Electrodynamics*, emphasizing the importance of his synthesis by alluding to Newton's *Philosophiæ Naturalis Principia Mathematica*.

Oersted continued to experiment and publish writings in daily newspapers. He made a famous contribution to chemistry by producing aluminium for the first time. Throughout his life, he remained involved in the field of philosophy, the source of inspiration for the truly groundbreaking experiment that resulted in a cascade of new scientific discoveries which underpin current technology and the societies it supports.

The universe as a quantum computer

When, centuries ago, the mechanical clock was the most complex device known to man, the universe was depicted as an enormous time-piece. There was even serious debate among experts as to whether God is obliged to occasionally wind up the mechanism.

Today we deal every day with devices that are increasingly complicated, but most of us don't imagine the universe as a giant iPhone or some other "wonder gadget" of our time. But this isn't true of some contemporary scientists. Seth Lloyd, a professor of physics at MIT, recently became famous for reimagining the universe as a giant quantum computer in a book entitled *Programming the Universe: A Quantum Computer Scientist Takes on the Cosmos* (Vintage 2007).

Will computers continue to improve?

Today's users of computers know that every few months they can get a device with much higher specs for the price of their old computer. In the last half-century, the processing speed of computers has doubled approximately every year and a half. This phenomenon in the development of the computer industry was observed in the 1960s by the former Intel manager Gordon Moore, and therefore bears the name "Moore's law". This law of course isn't a law of nature; it describes the inventiveness of computer engineers in designing smaller and faster microchips every eighteen months and doubling their processing speed. Every fifteen years processing doubles ten times in speed, the rate increasing by a factor of thousand. Today's computers are a billion times faster than those of half a century ago.

Searching for ever-new improvements is by no means a simple task, since the ever smaller wiring, transistors and other related electronic elements are getting ever harder to control. That is why

Moore's law is proclaimed dead whenever a technical issue arises that seems unsolvable at that particular moment. But so far every time this has happened a scientist or engineer has found a way around the problem.

If the engineers manage to keep up with Moore's law of technical progress, the basic microchip elements responsible for actual computing will diminish to the size of atoms in approximately forty years. The minimization process will then reach a limit that even the most brilliant engineers won't be able to overcome. Building-units smaller than atoms simply don't exist on Earth or anywhere else in the universe.

But the physical limit of size is not the only factor that influences a computer's speed and performance. The world of atoms follows different rules of behavior to the one we live in. In the microcosmos of atoms, a particle can exist in two places at the same time, travel on different routes simultaneously and perform other magic tricks described in detail by quantum physics. And by the time computer building-units reach an atomic level they too will be able to use the extraordinary logic of quantum reality, reaching even greater speeds in processing information. Today, the first basic quantum computers, consisting of only a few atoms, are being tested. Seth Lloyd, the author of *Programming the Universe*, is one of the most prominent experts in this futuristic field.

The greatest computer in the universe

In his book, Lloyd questions the final capacity limits quantum physics would appear to put on individual computers, provided that all the obstacles are reduced to laws of nature, disregarding any other technical issues. He imagines "the ultimate laptop" as an object one kilogram in weight and one litre in volume, similar in size to today's laptops, but with the added extraordinary feature of using each atom in its structure for computing.

Lloyd wants to assess the upper limits of computing capacities in a quantum computer. One of the operational speed limits he deals with is energy supply. The more energy a data-bearing particle contains, the faster it moves and the faster it calculates. Lloyd estimates the upper level of energy theoretically available for this type of computer by using Einstein's famous formula $E=mc^2$.

He figures that the ultimate laptop would deploy the same amount of energy released by a large nuclear bomb. And if this computer were someday actually started up, its interior would in fact look like an atomic explosion. According to Lloyd "the ultimate laptop looks like a small piece of the Big Bang."

He calculated that the ultimate laptop would be capable of performing ten million billion billion billion billion billion calculations per second. This corresponds to the ultimate upper limit of Moore's law. If Moore's law remains unbroken, Lloyd's ultimate laptop will become commercially available by the year 2200.

In the same way Lloyd determined the computing capacities of the ultimate laptop, he can assess the computing capacities of the entire universe if it is treated as a giant quantum computer. What would happen if we could use all the material and energy in space for performing a calculation? What would be the actual capacities of this kind of machine? As with the ultimate laptop, the main obstacle here is energy. Energy supplies in space have been determined fairly precisely by a number of astronomers, even though it seems that most of the energy is stored in very exotic forms such as dark matter and dark energy. But for Lloyd's estimation this isn't all that important.

Lloyd figured out that a computer running on all the energy of the universe would perform 100,000 googols of operations per second. (Googol is an enormous number, written with 1, followed by one hundred zeros. According to a famous anecdote, the founders of Google wanted to name their company after this number, but one of them made a mistake while filling in a document at the administrative office.) In the 14 billion years of its existence, this cosmic computer could have performed 10,000 billion billion googol operations.

No difference between simulation and reality

Besides determining the differences between quantum computers and those we use today, Lloyd wonders if a simulation of natural processes calculated at an atomic level would differ at all from the actual processes observed in nature. He finds no differences between real natural processes and a simulated occurrence calculated by a quantum computer. This not only means that a quantum computer is an ideal simulator of natural phenomena and the evolution of space, but also that space itself might really be such a computer.

But what does the universe actually do, if it *is* a giant quantum computer? Lloyd's answer doesn't come as a surprise. The universe calculates its own evolution. The data in this cosmic computer is stored in reciprocal data-sharing arrangements and other characteristics shared by the atomic and subatomic particles filling space. According to this analogy, collisions and other interactions between particles are nothing more than calculations performed by the universe. Every time two particles collide and change direction, for instance, the universe has done a simple calculation.

The fact that this giant cosmic computer calculates its own evolution means that the future of the universe can't be predicted accurately. For a detailed simulation of the universe, we would need a quantum computer that would be an exact copy of the real universe and would also calculate its own progress in time. But this wouldn't really get us very far.

Calculating the weather

The events that occurred at the peak of the Crimean War in the middle of the nineteenth century represent an important milestone in the history of weather forecasting. French and British fleets were caught in a violent storm on the Black Sea on November 14, 1854, which caused significant damage to a number of their choicest vessels. Reacting to the devastation caused by the hurricane as they followed the war's progress, the general public started wondering whether storms might be predicted and their terrible effects diminished.

The French emperor Napoleon III turned to one of the most respected scientists of the time, the man who accurately predicted the existence of the then unknown planet Neptune by calculating the discrepancies in Uranus's orbit. He was faced with a simple question: if science was able to predict faraway events in space years before they actually occurred, wouldn't it be possible to forecast weather conditions here on Earth at least a few days in advance?

Urbain Jean Joseph Le Verrier tackled the problem methodically. He began by collecting as much information as he could about the weather conditions between November 10-16, 1854, when the storm gathered over the Crimean peninsula. He used the data to draw up a series of maps, connecting dots where identical air pressure had been recorded. When he studied the temporal sequencing, it became apparent that near the eye of the cyclone there were marked drops in air pressure. This meant that drawing meteorological charts simultaneously with weather changes could predict the formation and movement of big storms.

Soon after his discovery, weather stations were built all across Europe, and incorporated into a network by the use of telegraphs. The storm forecasting service in England was led by vice admiral

Robert FitzRoy, best known as the captain of HMS Beagle on Charles Darwin's famous voyage a few decades earlier.

But the chief problem faced by early methods of weather forecasting was that they were based on observations and experience and not united by principles of physics. It was generally known, for instance, that a rapid decrease in air pressure indicates bad weather, but these observations weren't supported by equations that would enable scientists to calculate future weather conditions accurately.

Despite the unreliability of early weather forecasts, they became widely popular, partly due to their regular publication in daily newspapers. In serious scientific circles though, they were still somewhat of a laughing stock. FitzRoy was hurt by the scientists' lack of trust in weather forecasting, which allegedly led him to slice his throat in the bathtub with a razor blade and bleed to death on April 30, 1865.

After the death of its director a board of officials found the predictions of the British weather forecasting service unsatisfactory, and closed the program down.

How meteorology turned into arithmetic

The next milestone in the development of meteorology took place in 1904, when a Norwegian physicist called Vilhelm Bjerknes found a way of bolstering weather forecasting with calculations. In a visionary paper, he showed how weather forecasting could become a serious computational science.

He described his approach in terms of mathematical physics, stemming from known initial terms and equations which would enable one to calculate progression in time. With a good enough knowledge of the current state of the atmosphere, it should be theoretically possible to predict the next few days' weather with reasonable exactitude. Unfortunately, in a time without computers these calculations couldn't have been done quickly enough to provide any useful results.

The key idea in Bjerknes's approach was to divide the atmosphere into small units, atoms of weather that would change in time. He split the atmosphere into a grid, working on the assumption that the range of basic information available – temperature, humidity, pressure and wind_ would be the same in each atmospheric unit on the grid.

The same approach is used by meteorologists today, as they predict the weather more than a week in advance with the use of supercomputers. The model works by applying physical equations to data on the state of each individual point in the atmosphere and its surroundings, and calculating its progression in time.

These units for calculating weather forecasts are significant in size. They cover up to tens of square kilometers in an atmospheric map which covers the entire surface of the Earth. The map is used to calculate weather forecasts for a small area and more than a week in advance. The time frame varies from a few minutes to half of an hour, depending on the choice of the network and its resolution.

Bjerknes introduced medical vocabulary to the meteorological jargon: according to him, weather prognosis had to be based on an accurate diagnosis of the current condition. Without quality information on weather conditions in the present, it would be impossible to predict accurately how they would change.

Weather fronts

As a child, Vilhelm Bjerknes helped his father Carl, a professor at Oslo University, as he conducted scientific experiments. After graduating in physics and mathematics, he travelled first to Paris in 1889 to attend classes given by Jules Henri Poincaré, and later to Bonn, where he became Heinrich Hertz's assistant and collaborator. Together they conducted research that eventually led to the invention of radio. In 1892, he returned to Norway to complete his PhD and was later offered a lecturing position in Stockholm, where he developed his mathematical methods for describing atmospheric and oceanic phenomena.

In 1905 he presented his ideas in the US, and impressed the representatives of the Carnegie Institution so much that they offered to finance his research until 1941. After a few years at Oslo University, he was invited to join the University of Leipzig. In 1917 he returned to Norway and founded Bergen University's Geophysical Institute. He was assisted by his son Jacob, a Swedish meteorologist Tor Bergeron and his son's schoolmate Halvor Solberg.

Over the next few years, the four scientists came up with a theory that weather activities are concentrated at levels relatively close to the ground, where warm and cold air collide. These areas of increased weather activity were named after war battlefields: fronts. The team accurately described cyclone formation and their influence on the movement of weather fronts. Their innovative approach represents a key stage in the development of modern meteorology.

Fear of famine leads to weather observation stations

During World War I, Norway had difficulties providing enough food for its citizens. Their harvests yielded less than half their usual wheat supplies, and the prices of transatlantic imports were rising. They needed to come up with a way to boost their national agriculture.

At the beginning of 1918, Sweden introduced a special telephone number for farmers to call to receive the daily weather forecast. There was no such initiative in Norway. This spurred Bjerknes to act and he launched an initiative to help farmers provide larger food supplies by upgrading public weather forecasting.

Before the war, Norway received telegrams about the weather conditions in Ireland, the UK and the Faroe Islands, the usual sources of weather interference. But during the war, this information was no longer publicly available, and Norway had to establish its own weather stations, which often overlapped with the national submarine detection network.

To meet this need Bjerknes introduced more than data collection, he worked out a way of processing the collected information rapidly and providing a daily weather prognosis, in a technique now identified with the Bergen school of meteorology. Each morning around 8 o'clock, his team gathered in the attic of Bjerknes's house, collected telegrams with readings from weather stations across the country and then drew up weather charts. At about 9.30 a.m. they issued their daily forecast, making the lives of farmers and fishermen a little easier. When new data arrived, they updated their forecast as the day went on and made their prognosis more accurate.

LIFE

Thank you for not following our recommendations

In the spring of 1941 somebody knocked on the door of a cottage in the Tyrolean Alps north of Bolzano, home to a young bohemian poet Lucy Ramberg and her young son Mario. Soldiers were at the door. Even though the soldiers spoke German, the boy knew very well that they were members of the SS who had come to take his mother away, and that he might never see her again. His mother was arrested for anti-fascist activism and sent to a concentration camp, most probably Dachau, as a political prisoner.

From heaven to hell

Mario was born in October 1937 in Verona, out of a passionate love affair between his mother Lucy and an Italian air force officer, Luciano Capecchi. Lucy's father was a reputable archeologist, her mother a painter, and she grew up in Florence together with her two brothers. The family lived in a villa with a spacious garden and an array of servants, and the children were educated at home by private tutors. Lucy was fluent in no less than six languages. Her brothers grew up to become physicists, but she took after her mother and developed a passion for art. Before the rise of Fascism and Nazism, Lucy attended the Sorbonne where she taught languages and literature.

After the affair with the Italian officer, and her subsequent pregnancy, she quickly realized that marriage to Lucian would not be a happy one, so she turned down his proposal and moved with her son to Tyrol. After a few years in seclusion, she was tracked down by the army.

Following his mother's arrest, the three-and-a-half-year-old Mario was left by himself at a nearby farm, in the care of Lucy's acquaintances. Lucy had suspected that she would be arrested sooner or later, so she sold all her belongings and gave the money to nearby peasants, asking them to take care of her son as best as they could if she was taken.

Little Mario spent the next year with the peasant family as one of the children. He remembers living a very simple life there. Most of the family's time was spent working in the fields, trying to get by. They ground their own flour and made wine from their own grapes. It was during World War II, when the allies had already landed in the south of Italy and were making their way through to the north. American and British aircrafts flew regularly over the farm, and at one time, for no apparent reason, started shooting at peasants who were working in the field. Mario was shot in the leg, but fortunately suffered only a minor wound that soon healed.

A five-year-old homeless boy

After a year at the farm, Mario suddenly fell out of grace with the family and was no longer welcome. It is not completely clear whether he was tracked down by his father, or whether the money his mother left for him simply ran out. In either case, he was put out on the streets and left to his own devices. Not yet five years old, he spent the next few years of his childhood wandering the streets of northern Italy, hanging out with other homeless children, joining street gangs, ending up in numberless orphanages. He remembers spending a few weeks with his father, who turned out to be so violent that Mario soon realized he was better off taking care of himself than suffering such abuse. His father later died during an air raid.

Mario somehow managed to survive the war despite the suffering, cruelty and lack of food. He remembers being constantly hungry and witnessing many terrifying incidents that haunted him in nightmares for years to come. When he was eight years old, he spent an entire year in a hospital in Reggio Emilia, most probably suffering from typhoid, as he realized later on recalling his symptoms. He remembers the conditions in the children's hospital as being only slightly better than those on the streets. There were no sheets on the beds and the patients were fed chicory coffee and old bread crusts.

All the children suffered from the same basic symptoms. They woke up feeling slightly better in the morning, Sister Maria took their temperatures and promised that they could leave soon if they stayed one day without fever. But during the day their fevers raised so dramatically that they became delirious. The chances of a cure in this particular hospital, which served more as a refuge center than a health institution, were very slim.

But again, Mario somehow managed to survive, and on his ninth birthday a woman stepped through the hospital door and asked for him. It was his mother Lucy. After five years in a concentration camp, she had changed so much that even her own son didn't recognize her. She had been looking for Mario ever since the liberation in spring 1945 and finally found him after eighteen months of searching.

Mario vividly recalls that his mother brought him a new set of clothes and that they travelled together to Rome to take care of some legal formalities regarding their documents. In Rome, Mario took his first bath in six years. They then travelled to Naples and from there they took a boat to America, to stay with Lucy's younger brother, who sent them money for the tickets.

The land of opportunity

Mario never found out anything from his mother about her time in the camp, though he often tried to get her to talk about it. Lucy, who lived to be more than eighty years old, never fully recovered from the horrors of the war. Mario was taken to a number of psychologists and psychiatrists by his uncle and aunt to cure him of his nightmares, but what benefited him most was the loving environment he now grew up in.

Mario's uncle, Edward Ramberg, was a reputable physicist and one of the co-inventors of the electronic microscope and television. It was most probably he who introduced the sciences to Mario, who two decades later became a PhD student at Harvard, in the laboratory of James Watson, the co-discoverer of the molecular structure of DNA. Watson's mentorship and his approach to research were important influences in Mario's later scientific career. Among other things, Watson encouraged his students not to lose time on insignificant questions which would only yield insignificant answers.

Even though Mario had an opportunity to conduct research at one of the world's most prestigious universities, he decided not to follow this path in Boston, but moved to Utah instead. Despite his colleagues' advice, Mario was troubled by the overly competitive environment in which everyone strove constantly to prove themselves with new results and publications. He wanted to look into more fundamental issues and for that he needed more time and peace. The offer that came from Utah University was just right for this.

A risky venture pays off

In 1980 Mario filed for grants from the US National Health Institute, which offers federal funds for medical research projects. He applied with three different projects. Two of them rested on solid grounds, had a clear timeline, offered intermittent results, along with predictable progress and a guaranteed outcome. The third one was highly speculative. Its aim was to perform a controlled change in a single mouse gene, a goal which at the beginning of the 1980s seemed complete science fiction.

The reviewers advised him to concentrate on the first two projects and let go of the third, since it qualified as a "mission impossible." He did receive the funding he applied for, but on the condition that he gave priority to the projects guaranteeing solid results.

Mario ignored the reviewers' advice and invested all his money and energy in changing the mouse gene. He was consciously risking his career as scientific researcher, knowing he would most probably not receive any more funding in the event of failure.

But his persistence paid off. In 2007 Mario Capecchi, together with Martin Evans and Oliver Smithies, received a Nobel Prize in Medicine for work on mouse genes. The reviewers from the National Health Institute had previously acknowledged that they were very pleased he decided not to follow their recommendation. Today, laboratories all over the world hold millions of these genetically modified or "knockout" mice. They have proved essential to many new genetic discoveries and in finding new cures for a number of terrible diseases.

The rocky road to finding a cure for cholesterol

In the cafeteria at the Japanese University Tohoku, a bell rang every evening at about eight o'clock, and whoever came running quickly enough received the day's leftovers. Many poor students lived on these free evening meals during their study years. Among them was a future biochemist Akira Endo, who like the others often waited for the cafeteria bell and ran to get some of the free food. Even though he later spent many years conducting research at Sankyo, a pharmaceutical company, working hard to discover one of the most profitable medicines in the entire pharmaceutical industry, he received no additional compensation for his success beyond his usual pay and pension.

After concluding his studies, Endo got a job at Sankyo where his initial research work consisted of isolating certain constituent parts or products of yeast organisms that would prove commercially viable. During that time in Japan, every student had to complete at least five years' work as a researcher in a relevant industry before they could obtain their doctorate. By the time he submitted his dissertation, Endo had already written no less than 15 research papers. His hard work and findings, which proved extremely lucrative for the company, brought Endo the chance to do post-doctoral training in the US in 1966. While in New York, he decided to switch his area of research to the biochemistry of lipids, commonly known as fat and cholesterol.

Good and bad cholesterol

Cholesterol is one of the key components of cellular membranes with important functions in the digestion of fat and the formation of hormones including estrogen and testosterone. 93% of all body cholesterol is stored in our cells, with only 7% in the blood, and is ingested with food or produced by the liver.

Cholesterol is carried around with help from two sorts of molecule called lipoproteins. LDL (low density lipoproteins) carry cholesterol from the liver to other parts of the body, while HDL (high density lipoproteins) return any surplus of cholesterol from the arteries, where it can cause a number of problems, back to the liver. Of course, both types of lipoprotein are essential for maintaining normal bodily functions, yet in today's vocabulary of heightened nutritional awareness, the HDL type has been labelled "good cholesterol" for its vein-cleansing properties, whereas LDL has been branded as "bad cholesterol". If our blood contains an increased amount of LDL particles, they begin sticking to the walls of the arteries and cause problems by forming plaque, which in the worst-case scenarios cause our veins to clog, raising the likelihood of a heart attack or stroke.

Searching for a magical substance among thousands of samples

After returning to Japan, Akira Endo set his mind to finding a substance that would prevent the formation of new cholesterol in the body. It had been known since the 1960s that the enzyme HMGCR (HMG-CoA-reductase) was responsible for cholesterol regulation. Endo needed to find a substance that would inhibit this enzyme, which plays a central role in the production of cholesterol in the liver. But where should he look? A few decades earlier, a number of effective antibiotics were discovered among microbes. Endo was certain that he would find a cholesterol-regulating medicine by thoroughly researching the world of microorganisms.

He embarked on a colossal project, studying thousands of samples of different soils, fungi and other organisms that might provide the right medicinal properties. He worked out a procedure for combining rat liver extract with a radioactive marker that enabled him to measure the amount of cholesterol produced by a correlated increase in radioactivity. He was soon able to confirm the existence of certain substances that inhibited cholesterol production, but it proved difficult to isolate them and rule out unwanted side effects.

At the beginning of the 1970s, after looking at thousands of samples, Endo and his team discovered four potential candidates among the substances that decreased cholesterol levels in the blood, two of which appeared especially promising. One of the samples was isolated from a parasite found on rice and cucumbers, called Citrinin, but unfortunately proved to be toxic. The other originated from microorganisms found on rice bought in the 1950s in Kyoto.

After two and a half years, the team tested 6392 samples, of which they found only one was suitable for animal testing. They named it Compactin. Unfortunately, rat tests showed no long-term effect in lowering cholesterol levels, even though it inhibited the functions of the HMGCR hormone responsible for the synthesis of cholesterol. It appeared that the mechanism regulating cholesterol levels in the rats' blood was so powerful that it reacted to the reduction of HMGCR by triggering additional mechanisms which brought cholesterol levels back up.

It already seemed that further research on Compactin would be abandoned when in January 1976 Endo met an old colleague in the cafeteria who had just finished experimenting on chickens. Since the chickens were headed for the slaughterhouse anyway, Endo decided to test his new drug on them. This time the results were very encouraging. Cholesterol levels in the test group dropped by half and no harmful side effects were observed.

Endo's first patient

In 1977 Endo was contacted by a doctor treating a girl with a severe case of hypercholesterolemia who had already suffered great problems from excessive cholesterol accumulation. Since the company couldn't officially allow the medicine to be administered to the patient, Endo endured a long moral struggle before finally travelling to Osaka with a dose containing fifty grams of Compactin and handing it to the doctor.

The eighteen-year-old girl became the first patient to be treated with one of the cholesterol-lowering drugs now known as statins. Her level of cholesterol soon dropped by 20%, but three weeks into her treatment she developed severe muscle pain. The doctors adjusted her dosage and lowered the medication eight other patients were receiving. Cholesterol levels dropped by an average of 30% in the test group, a result that was good enough to put the medication through to the first phase of full clinical trials.

Endo then decided to leave Sankyo and accepted a teaching position at Tokyo University. This change of employment was unheard of at the time in Japan's conservative corporate culture, but Endo hoped he would be treated with understanding given his contributions to the company. He was wrong. After twenty-one years and eight months of dedicated work, he had to clear his office, pack his things and leave without any help from his colleagues.

Pharmaceutical companies and their astronomical profits

The Sankyo company itself was very naïve in consenting to share all its results with researchers from Merck, the firm that offered to help develop the final product. In the end, Merck cheated Sankyo out of the deal and patented the winning version of the statin themselves.

While doing research at the university, Endo discovered a new form of statin in another variety of rice and proceeded to file a patent. The Merck researchers contacted him immediately and asked for a sample. Endo reluctantly obliged. Even though Endo was the first to patent the substance, Merck insisted that they had invented it a few months earlier but circumstances had prevented them from filing a patent. Merck managed to convince the US Patent Office they indeed were first and in 1987 they launched exactly this substance under the brand name Lovastatin.

Akira Endo, today in his eighties, lives in Tokyo and still enjoys hiking with his friends in the neighboring hills. A few years ago he said in an interview that his personal physician found his cholesterol levels slightly increased and told him: "Don't worry. I know a very good drug."

Cell police

A long, long time ago when there was no vegetation and no animals roaming our planet, its sole occupants consisted of simple single-celled organisms that developed an ingenious mechanism to protect themselves from the attacks of vicious viruses that were already causing major epidemics. Scientists have only recently discovered that the very same mechanism, developed by single-celled organisms for fighting viruses, is still active in most living beings on Earth, including humans. A thorough understanding of this mechanism should enable us to treat effectively a number of serious conditions. So it didn't come as a surprise when the 2006 Nobel Prize in Medicine was awarded to the scientists who discovered the secrets of this prehistoric, yet still very efficient cellular mechanism possessed by the majority of living beings.

Like a highly regulated country

A living cell is an exceedingly complicated instrument containing a number of precisely executed and carefully controlled chemical processes. It could be compared to a country with strict regulations. The central governmental headquarters holds a hefty volume of the penal code listing and defining all the rules of production. This is the only copy and it can never leave the central office. Since the inhabitants of the country need specific instructions on how to produce anything, the government employs numerous copyists who incessantly transcribe instructions for making specific products and send them throughout the country. When a transcript reaches its addressee, they immediately get to work. If the addressee is an engineer, he or she performs a requested task according to strict specifications.

In the above analogy of an over-regulated state, we have a rough sketch of the way a living cell functions. The unitary code holding all the information for a well-guided state is of course a metaphor for the genetic transcription that is safely stored in the DNA molecule and never leaves the cell's nucleus. The transcription of instructions on the production of individual parts stand for messenger RNA molecules, which are responsible for transmitting information from the nucleus to other parts of the cell, while the faithful copyists of the centrally stored code represent RNA polymerase enzymes. The 2006 Nobel Prize in Chemistry was awarded for studies on this enzyme, which is responsible for transcribing DNA into RNA.

"Transcriptions," that is, messenger RNA molecules, are allowed to break away from the cell nucleus and flow freely in the cell. As for the engineer mentioned above, who receives schematics and follows them exactly, in the real world of cells, these workers are called ribosomes. Their task is to use the information inscribed in the messenger RNA for building a specific protein which will perform specific tasks within the cell. This is how nearly every living cell on Earth works.

Cell terrorists

There are however many downsides to this strictly regulated country. Evil terrorists can smuggle in their own notes, instructions that aren't copied from the headquarters' penal code. When this "terrorist blueprint" reaches its recipient, the engineer isn't able to tell whether the note came from a credible source. Nothing prevents the engineer from following the terrorist's dictations, thus inadvertently complying with the demands of the enemy by producing a potentially deadly murder weapon. These evil terrorists that smuggle into the cell and sabotage its transmission mechanisms are called viruses. They are extremely mischievous and well adapted to this kind of piracy.

Anti-terrorist units

Fortunately, cells evolved a method for fighting these attacks. When terrorists enter a cell, the first thing they attempt to do is multiply, or in terms of our metaphor, produce as many recipes as possible to make their murder weapons and spread them throughout the cell. As they multiply, while creating these damaging recipes, the viruses stay inside the double-stranded RNA. We could say that for a certain amount of time, the original recipe and its copy remain bound together. And precisely at that time, while the authentic instruction and its copy are still one, the cell can identify the recipe as potentially dangerous and dispatch a special defense unit. This special anti-terrorist force not only sets out to destroy the renegade orders, it disintegrates all the recipes in the cell that contain genuine instructions.

The official name of this anti-terrorist unit, trained to detect and destroy rogue messages, is RNA interference or RNAi. For their significant contribution to understanding how the cells' virus defense mechanism works, and how it is also helps regulate certain genes, American scientists Andrew Fire and Craig C. Mello received the Nobel Prize in Medicine.

The Petunia mystery

As often happens with important scientific discoveries, the first encounter with the RNAi mechanism was sheer coincidence. In 1986, a small biotechnology firm in California set out to create a unique flower. They decided to make a petunia with vividly purple-colored blossoms. They located the plant's purple gene and tried to make it more prominent by simply adding instructions to produce more purple to the petunia's DNA code.

To everyone's surprise, the flowers of the newly created genetically modified petunia, instead of blooming in a stronger purple, came out pure white. The researchers initially thought something must have gone wrong, but careful scrutiny of the procedure revealed no mistake. Numerous brainiacs spent more than a decade trying to find out why the petunias became white instead of purple.

Full-blown terrorist alert

Today we know that the party responsible for intercepting the transmission of new genes into petunias was the RNAi mechanism. It made sure that the artificially included new genes in the cell were interpreted as an invading virus and immediately proceeded to destroy all the instructions that were identical to those the scientists had tried putting in the cell. Since the newly introduced recipes were identical to those produced by the cell, the cell's anti-terrorist unit destroyed both the artificially induced purple-signals as well as the authentic purple codes produced by the copyists in the cell's nucleus. The cell's defense unit thus successfully eliminated all the cell's recipes for producing purple petals, causing the petunia to blossom white.

For now, the RNAi method offers a strong tool in scientific research of individual genes, but it may soon also provide an effective means of diagnosing and treating a number of serious health conditions. A cell's anti-terrorist units can be conditioned to eliminate specific recipes containing instructions for the production of particular proteins. A number of diseases result when cells overproduce a particular ingredient. Scientists are already testing new ways of using the RNAi mechanism to treat various forms of hepatitis, Huntington's disease, AIDS and certain kinds of cancer.

The study of digestion through a hole in the stomach

On June 6, 1822 a hunting accident occurred on Mackinac Island near the Canadian-US border. A musket accidentally went off at the trading post of the American Fur Company and a young employee was shot in the chest at close range. Alexis St Martin, the victim, was brought to the ground by the impact of the shot, the pellets penetrating his chest and setting fire to his shirt. All the bystanders were convinced that the young man was beyond help.

Fur trader's accident

As luck would have it, the shot was overheard by an army doctor, William Beaumont, who was working nearby. He attended to the victim in a few minutes by giving him first aid. He cleaned the wound and stopped the bleeding, but he too was not optimistic about St Martin's fate. The injuries were too severe. The wound on the left side of his chest was the size of a tennis ball: St Martin had fractured ribs, and had suffered injuries to his lungs and abdomen. The shot had blown off fragments of his muscles and skin. The doctor estimated that St Martin wouldn't live more than thirty-six hours. But fortunately he was wrong.

For the first few days after the accident it seemed that St. Martin's fate was sealed. His injuries included a hole in his stomach, through which all the food he ingested seeped out. It took a long seventeen days for his condition to improve, so he could eat and digest normally. After a month, St. Martin's health stabilized, but Dr Beaumont still needed to attend him with care and clean his wound regularly. It took several operations to remove the pieces of clothing that had been pushed inside his body by the pellets.

The hole in the stomach remains open

The story itself would be nothing special if it hadn't turned Alexis St. Martin into one of the most famous "guinea pigs" in the history of medicine. The wound in his chest eventually healed completely, but not in an entirely normal way. The accident caused his tissues to grow back in a way that formed a direct connection between his stomach and the outside. This "hole in the stomach", called a "permanent stomach fistula" brought St Martin global fame. (Today, similar gastric fistulas are surgically inserted into testing animals for the needs of digestion research.)

After ten months of treatment, the head of staff decided to send St Martin home, since there was not much more that could be done for the patient. But his home was more than two thousand kilometers away, a distance too great for the weakened patient. So Dr Beaumont invited him to stay at his house for the time being. Eventually St Martin made complete recovery and in April 1824, less than two years after the accident, Dr Beaumont employed him as his housekeeper, tasked with chopping firewood and other "masculine chores" around the house.

But even after his troubles were over, St Martin still needed to tend to his hole in the stomach, covering it and changing its bandages to prevent his most recent meal from pouring out. Dr Beaumont's reasons for never closing the aperture surgically remain unknown. Perhaps it wasn't possible initially, and soon proved too good an opportunity to observe in real time what goes on in the human stomach. In the following years, Dr Beaumont was able to solve many unanswered questions about digestion thanks to St Martin's stomach fistula.

How does digestion work?

Dr Beaumont's approach to researching St Martin's stomach was extremely methodical. Through the fistula of his human guinea pig (whom he paid extra for putting up with these unpleasant experiments), he inserted different foods and monitored gastric activity. He tried out all kinds of meat, fruits and vegetables. Poor St Martin often suffered from indigestion due to the bizarre mixtures in his stomach, and soon the financial incentive he was offered could no longer compensate for all his troubles.

Fed up of being scarified in the name of science, he fled to Canada, married, and fathered six children. But Dr Beaumont located him after a few years' search through the fur trading company St Martin was working for, and persuaded him to move back with his family to continue the experiments. With these new tests, Beaumont gathered enough information to write a book on human digestion. The book was published in 1833, entitled *Experiments and Observations on the Gastric Juice and the Physiology of Digestion*. The work brought Beaumont worldwide fame and made him an illustrious name in the history of medicine.

In his book, Dr Beaumont describes over 200 experiments he conducted ion his patient's famous stomach. He proved irrefutably that digestion is a chemical process, putting an end to a lengthy controversy stretching back to the beginnings of medical science. The nutrition tables he published in his book served as key references for a healthy diet for nearly a century. Among other things, his experiments proved that meat is digested faster than vegetables, which brought sneers from a group of doctors who were early supporters of vegetarianism. To refute his claims, they tried to rent St. Martin's stomach themselves to test out their vegetarian principles "scientifically", but could no longer convince St. Martin to co-operate with such experiments.

The ethics of scientific research

A debate on the ethics of experimenting on live humans opened only in 1834, when Beaumont requested state funds for financing further research. The controversy surrounding his research methods had already spread by then, judging by an anecdote from a murder trial that featured the famous doctor. In 1840, while Beaumont was working in St Louis, among his patients was the editor of a local newspaper who was struck on the head with a metal stake by a politician who disapproved of his editorial column.

Dr Beaumont attempted to save the editor's life by drilling a small hole in his skull to release pressure on the brain. This technique was later disputed in court, when the politician's lawyers tried attributing the editor's cause of death to incorrect treatment rather than a blow from a metal stake. The counsel for defense argued that the fascination which led Dr Beaumont's to explore the hole in a patient's stomach, out of pure scientific curiosity, had now led him to look inside a man's head by drilling into his skull. This proved a successful defense tactic leading to a five-hundred--dollar settlement of the case.

Despite all the bizarre adventures he endured, St Martin died at the venerable age of eighty-six, on June 24, 1880 in Quebec. After his death, his relatives refused all requests by doctors to perform an autopsy on the famous patient's body. He was buried in an unusually deep grave and covered with a heap of rocks to keep his remains safe from the harassments of further researchers.

How a surfer won the Nobel Prize

> The sun had been hot that day in Mendocino County. It was May. A dry wind had come out of the east, and nobody knew how hot it had been until, around sunset, the wind stopped. I drove up from Berkeley through Cloverdale headed to Anderson Valley. ... My little silver Honda's front tires pulled us through the mountains. My hands felt the road and the turns. My mind drifted back to the lab. ... Tonight I am cooking. The enzymes and the chemicals I have at Cetus are my ingredients. I am a big kid with a new car and a full tank of gas. I have shoes that fit. I have a woman sleeping next to me and an exciting problem, a big one that is in the wind.
>
> "Holy shit!" I hissed, and left off the accelerator. The car coasted into a downhill turn. I pulled off. A giant buckeye stuck out from the hill. It rubbed against the window where Jennifer, my girlfriend and co-worker, was asleep, and she stirred. I found an envelope and a pencil in the glove compartment. Jennifer wanted to get moving. I told her something incredible had just occurred to me. She yawned and leaned against the window to go back to sleep. We were at mile marker 46.58 on Highway 128 ... I had just solved two major problems in DNA chemistry. ... I'll be famous.

This is how biochemist Kary Mullis, in his autobiography *Dancing Naked in the Mind Field* (Random House, 1998), describes the unforgettable moment in 1983 when he was hit by what was probably the most important idea of his life, and for which he was awarded the Nobel Prize ten years later. He went on to invent a method for the artificial replication of a DNA molecule *in vitro,* outside the living cellular environment. Today, his method, officially named polymerase chain reaction or PCR, is indispensable in any biochemical lab facility, where it is physically present in the form of a small table device. Without these PCR devices, we wouldn't be in the 'era of biotechnology', as this fast-moving period of incredible success in the chemistry of living organisms is often called.

Nobody believes me

Of course a real story is not necessarily a happy story. When Mullis got the idea, he was convinced that the method he devised was so simple that it must had been thought of by someone before him who had tested it, proved it was useless and forgotten all about it. When he and Jennifer reached Anderson Valley, Mullis opened a bottle of cabernet and thought long and hard about possible faults in his idea, searching for reasons why it wouldn't work. Although he and Jennifer were on a weekend mini-break, he could hardly wait until Monday, when he would go to the library, bury himself in scientific journals and find out whether somebody had made the discovery before him.

By noon on Monday, it was already clear that the literature contained no writings on the subject. Yet his colleagues didn't seem very thrilled with the idea. Mullis attributes it to the fact that it was common for him to get similar crazy ideas almost every week, and everyone was fed up listening to him. But this he was absolutely certain was no ephemeral flash of wit that he would forget by the following week. During the first few months, the only person who believed in him was a good friend who had opened his own biotechnology firm; but even he is presumed to have supported Mullis out of opportunism, since if it did work the new technique would prove great for business.

This friend advised Mullis to quit his job, because nobody in the company believed in his theory. The friend told Mullis he should develop the idea, put a patent on it and get rich. Of course neither of the two biochemists, who were then busy weighing up their decisions for the future, could have imagined that only a few years later, Cetus, the giant company Mullis was working for, would buy the PCR method from Hoffmann-La Roche for three hundred million dollars.

How to copy molecules?

The main principle of the PCR method is best described by an analogy. Gene transcription in the form of double-stranded DNA can be compared to text printed in a book. When the book is closed, it's impossible to read, transcribe or photocopy its contents. The same goes for DNA. As long as the molecule stays double-stranded, the information it stores can't be copied or used in any other way. Only when it opens and the two chains break away from each other does the text become legible, and comparable to an open book. Yet double-stranded DNA outside a living cell may be 'opened' simply by briefly raising the temperature to about 95 degrees Celsius.

The PCR method enables us to copy specific segments of DNA when we know enough about the beginning and the end points of what is going to be transcribed. In terms of our analogy, this is like copying a section in a book by knowing only the first and the last sentences of the text. Let's suppose that we do know those sentences already, and it is the intervening text, perhaps many hundreds of pages long, that we need to find. By opening the book, and skimming through it until we find the corresponding sentences, we can identify, read or copy the passage we're looking for.

The counterparts to these initial and final sentences in the world of DNA molecules are short, artificially produced pieces from a single-stranded DNA molecule that stick to the designated spot in a molecule opened by heat treatment. When the temperature is brought back down, the two pieces attach themselves to the molecule and enable the corresponding enzymes to fill in the space between them. This copies the part of the DNA sequence between the first and last sentence in the genetic transcription, flagged by two sticky pieces of DNA.

A key factor in the procedure is that it involves a chain reaction, meaning that the same process is repeated several times in a row. With each repetition, the number of copies doubles, which theoretically gives us a million identical copies of a single DNA molecule after only twenty repetitions.

Will they call back?

On the October 13, 1993 Mullis was woken by a telephone call at 6:15 a.m. Based on his previous experience, he thought someone from Japan with no sense for time zones was trying to send him a fax. Rather than getting out of bed, he waited for the fax machine to take over the call. But instead it went to voice mail and Mullis was stirred from his sleep by hearing the words "the Nobel foundation". He jumped out of bed and picked up the phone, but the caller hung up at that very moment. This is how he remembers that October morning: "Great, I thought, I've missed the Nobel Prize call. Will they call back? Almost instantly the phone rang again. He had heard me just as he'd hung up. 'Congratulations, Dr Mullis. I am pleased to be able to announce to you that you have been awarded the Nobel Prize.' 'I'll take it!' I said. I knew they couldn't make you take it and I didn't want there to be any doubts."

The man warned him to expect a deluge of calls from reporters, but Mullis couldn't have anticipated the extent of what was about to hit him. When he got off the phone, he wanted to call his mother to South Carolina since it was her birthday and he knew the news would make her day. But as soon as he picked up the receiver, a journalist was already on the other end of the line. The phone didn't even have the chance to ring. After a few seconds of conversation, he put down the phone and picked it up again. It was another journalist informing him that he was coming over with a TV crew.

The surfer who got the Nobel Prize

Soon afterwards, a neighbor came to his door. Mullis told him he had just found out he was getting the Nobel Prize. The neighbor said: "I know. I heard it on the radio. Let's go surfing." A few other friends joined them. When they came out of the water after a morning wave-ride off the Californian coast, a camera crew was already waiting on the beach. The journalists and cameramen of course had no idea what Mullis looked like, and ended up asking another surfer how it felt to be a Nobel laureate. "Surfer Wins Nobel Prize" was a common headline that day.

When Mullis finally managed to call his mother that afternoon, he hoped that now he had got the award, she would finally stop sending him newspaper cut-outs about advances in DNA chemistry and start believing his expertise extended beyond the skills of popular science writing. He joked about his mother probably hoping that winning the Nobel Prize might get him a mention someday in *The Reader's Digest*.

In his acceptance speech in Sweden, Mullis didn't describe the method itself but rather talked about the feelings that came over him at the time of its discovery. Among other things, he mentioned that science to him meant having fun, and that the PCR method was a continuation of what he had done as a child. His intention hadn't been to revolutionize biochemistry; he viewed his invention as a tool he had created to help him conduct other experiments. But he admitted to having been very naïve. If he had realized what he was doing at the time, he said he would never have discovered the method.

The Neanderthal genome project

In August, 1856, Germany, a group of workers was cleaning out a cave near Düsseldorf, Germany. The cave was one of the many in the Neander Valley, named after a local poet and teacher, Joachim Neander. While throwing big chunks of clay out of the limestone cave, they came upon a few bones and the remains of a skull. At first, the workers were convinced the bones belonged to a cave bear, a common find in the area, but still they decided to show their discovery to a local teacher, Carl Fuhlrott. He soon realized the remains didn't belong to a bear but a human.

Because the bones had some unusual features, the teacher consulted Professor D. Schaafhausen from Bonn. Schaafhausen agreed the bones were very unusual and different from even "the most barbaric human races". Nobody could tell to whom the remains belonged, which naturally gave rise to a number of more or less hypothetical interpretations about their origins. Some were convinced the bones belonged to an ill and deformed modern human, maybe even a soldier who had fought against Napoleon and died in the Neander Valley. Others placed the remains in a more ancient historical timeframe, in the era when people still lived in caves.

It took quite some time for the scientists to agree that the remains belonged to a humanoid species that was closely related to modern humans and had existed in Europe until 28,000 years ago. The species was named Neanderthal, after its most famous place of discovery, even though 'Neanderthal' remains had already been found in various locations across Europe. Today we know that modern man lived next door to the Neanderthal for quite some time, and according to the latest genetic research the Neanderthal and homo-sapiens gene pools got mixed together, which turned out to be very useful for the latter strain, enabling it to adapt faster to an unfamiliar environment.

Genomes of extinct human ancestors

If we compare the entire genetic code, called a genome, of two unrelated living people today, the "letters" in their individual codes will differ approximately every three thousand times. The human genome contains 3.2 billion of these genetic letters, with the differences between two unrelated individuals amounting to approximately 3 million letters. For comparison: in comparing a human being and a chimpanzee, differences would occur in every hundredth letter of code.

Coincidental genome mutations multiply more or less proportionately with time, which means we can estimate when two creatures shared a common ancestor from the changes in their genomes. Humans share a common ancestor from about half a million years ago, while we share a common ancestor with the chimps from five million years ago. If there are ten times more differences in two sets of genomes, the two species shared an ancestor ten times further back in the past.

It is interesting to observe how the differences between those alive today became distributed across different parts of the world. The greatest genetic variability in humans is found among the peoples in Africa, even though this continent houses only a small proportion of the world's population. A part of this African variability is characteristic only of Africans and cannot be found anywhere else, which means that people in other parts of the world are genetically less diverse than living members of ancient African tribes such as the Pygmies and Bushmen. Researchers explain this as the result of a partial migration of the original population of modern humanity's ancestors from Africa to other continents. This means that all non-Africans today are descendants of ancestors who left Africa about 65,000 years ago. In short, we all come from Africa.

Swedish geneticist Svante Pääbo has long been developing techniques for analyzing the genomes of extinct humanoid species such as Neanderthals. He has been able to isolate enough genetic material from preserved bones to reconstruct more than half of the Neanderthal genome. Most of the remains containing sufficient DNA were retrieved from the Vindija cave in Croatia, near the Slovenian border.

Since the Neanderthals never inhabited Africa, Pääbo's analyses show that Europeans are genetically closer to Neanderthals than the descendants of people who never emigrated from Africa. This can only mean that those among our ancient ancestors who exchanged the African for the European continent sooner or later bumped into Neanderthals, and exchanged genetic material by mating. This genetic transfer apparently occurred soon after our predecessors left Africa, since the people who eventually colonized other continents already carried Neanderthal genes.

A Siberian girl's pinkie finger bone

In 2008 Russian archaeologists discovered a tiny bone in a South Siberian cave called Denisova - named after the hermit Dionisij who lived in it in the eighteenth century. The bone came from a pinkie finger of a child aged between five and seven, and had ended up in the cave more than 40,000 years earlier. The researchers managed to isolate a great deal of genetic material from the tiny bone, and the results of their analysis caused quite a sensation among scientists. What they found was that the genetic code of the bone was considerably different from both the human code and the Neanderthal code.

According to the results of the analysis, this newly discovered humanoid species shared a common ancestor with Neanderthals

640,000 years ago, while it shared a human ancestor 800,000 years ago. Further research showed that humans mated – or "exchanged genetic material" – with the newly discovered species, but the proof for this was found only in native inhabitants in Indonesia and Papua New Guinea.

Because the genetic material from this hitherto unknown humanoid species is still very scarce, scientists tend to be wary of drawing any conclusions, but a new picture of what was going on in Europe and Asia tens of thousands of years ago is slowly coming to light. If current research is correct, about 50,000 years ago Europe was inhabited by Neanderthals, while Denisovans – as this newly-discovered human-like species has been named – roamed East Asia.

When modern humans arrived from Africa some of them cross-bred with Neanderthals in the Middle East and inhabited Europe, while others went on to Asia, cross-bred with Denisovans and travelled further to other continents.

Recently published studies of genetic material from humans today from all around the world, and the reconstruction of genomes from extinct human-like species confirm that a multiple crossbreeding of genes occurred between the modern human and other humanoid species of the time. Even people that never left the African continent exchanged genes with those species that once lived there.

Crossbreeding with Neanderthals, Denisovans and other species that still haven't been discovered was apparently a great benefit to our ancestors. It increased their immunity to some diseases typical of Europe and Asia. Both extinct branches of humanity had lived for a long time in the territories where immigrating *Homo sapiens* humans first encountered them, and had therefore developed immunity to many viruses and other sources of disease which our human ancestors from Africa weren't lucky enough to gain. This is why the mixing of genes of which we have increasing proof enabled the children of some migrants to survive in their new and hostile environment.

The first intensive care unit

In the fall of 1952 Division 19 of the infectious diseases hospital in Bledgams, Copenhagen was packed to breaking point. Seventy beds were filled with severely ill children who had been attacked by the polio virus. A tube was surgically inserted into their windpipes, attached at the other end to a rubber balloon and a ventilator pumping air from an oxygen tank. Next to each bed sat a medical student whose job was to squeeze the balloon every few seconds, delivering an oxygen-enriched air supply to a child who couldn't breathe without this assistance.

The logistics of long-running manual balloon pump oxygenation

The students were relieved every six or eight hours, which meant that three to four times a day a new group of young men and women arrived at the hospital and sat continuously by the children to keep them alive. During the eight-hour shift they were allowed a ten-minute break every hour, and a mid-shift lunch break of half an hour. While attending, they spent hours pumping the balloons and reading stories to the younger children, trying to distract them from their grim reality.

This intensive care regime at the hospital lasted more than six months. Many young volunteers gave up after a few weeks of this strenuous and responsible work, which exhausted them physically and emotionally. Yet overall morale was still high and everyone did their best to help. Each individual child was assigned four or five students because seeing familiar faces every day proved much less stressful for the young patients.

Of course the students, most of them sophomores at the medical school and usually younger than twenty, weren't the only people taking care of the patients. The department employed a number of doctors and nurses who helped out whenever a complication arose and took care of supplies and equipment. Besides the oxygen tube, each child was equipped with a special receptacle for CO_2 that needed changing regularly.

During the polio epidemic, the intensive care unit at the Blegdams hospital housed 1500 volunteers, whose selfless work helped reduce the mortality rate of the most afflicted children from 90% to 25%. By 165,000 hours of work they saved more than a hundred children from certain death. The children wouldn't have survived without respiratory support.

Iron lungs proved unsuccessful

Denmark's polio epidemic of 1952 proved an important milestone in the history of medicine, making assisted respiration by means of catheter a primary aid to severely ill patients in intensive care units around the world. Prior to this, surgically inserted tubes were used only rarely during some operations.

In 1931 an American Philip Drinker had invented the "iron lung" for use in assisting respiration. The machine used negative pressure to expand the lungs rhythmically, ensuring a regular inflow of fresh air. It was used for patients whose breathing muscles were severely weakened by the virus, to keep them alive until they recovered. But iron lungs had many drawbacks and, among other things, made it difficult to keep the patient's airways open.

In September 1951, the University of Copenhagen hosted an international conference on polio that took place in a very optimistic atmosphere. Among the participants were Jonas Salk and Albert Sabin, the discoverers of the polio vaccine that was about to be released for general use. But it seems that one of the doctors at the symposium was an asymptomatic carrier of the virus, making the conference the epicenter of the deadly epidemic that swept the country a year later.

Copenhagen wasn't ready for this major polio outbreak, and the Blegdams hospital housed only one iron lung and six similar smaller devices. Because they had treated only ten cases requiring such treatment in recent years, the hospital chiefs didn't think they'd need more equipment. Despite the iron lung treatment, the mortality rate was still high at 80%. In the first three weeks of the epidemic, twenty-seven patients out of thirty-one died from respiratory failure, of which nineteen died within three days of being hospitalized.

Critical conditions lead to a revolution in the understanding of polio

By August 1952, the epidemics had reached alarming proportions. Thirty to fifty new patients were brought into the hospital every day, with between six and twelve in a serious condition. By the end of the year, 2722 polio patients were hospitalized, half of whom were partially or fully paralyzed, with 349 suffering from respiratory paralysis. The hospital management realized that buying more iron lungs would not solve the problem. They needed new ideas.

One of the doctors suggested consulting the hospital's anesthesiologists about the patients suffering from respiratory problems, since they had experiences supporting patients' breathing difficulties during complicated surgery.

Dr. Henry Cai Alexander Lassen, the technical director at Blegdams hospital, wasn't immediately convinced by his colleague's suggestion because of the prejudices he held against anesthesiologists. He believed they weren't real doctors but merely hospital technicians, responsible for administering the right doses of anesthetics; but in the absence of a better idea he eventually gave in.

On Saturday August 25 they called in the anesthesiologist Bjørn Ibsen MD who, after long deliberation, put forward a solution that did wonders not only for the treatment of polio, but marked a revolutionary breakthrough in the whole field of twentieth-century intensive care. In order for Dr Ibsen to get his ground-breaking suggestion put into practice, he needed to change his colleagues' views on some of the basic medical doctrines of the time regarding the cause of death in polio.

In the last stage of the disease patients usually develop high blood pressure, followed by fever soon followed by death. These symptoms were interpreted as primary features of the disease; it was believed that in the final stage the virus attacked some of the key brain centers controlling basic bodily functions. But Dr. Ibsen didn't agree with this theory.

After carefully analyzing the different stages of polio, he suggested that the actual cause of death wasn't the progression of the viral infection but poor oxygenation of the tissues, the result of respiratory failure. He came up with a radical solution: patients should have a tube surgically inserted into their airways in order to receive assisted breathing. This intervention would regenerate patients' tissues until they were strong enough to recover.

A successful experiment convinces the sceptics

The technical director of the hospital wasn't convinced but still gave his permission for an experiment to be conducted. On Monday August 27, 1952, Ibsen chose a severely ill twelve--year-old girl, Vivi, who had been admitted the previous day. All four of her limbs were paralyzed from polio, she had trouble breathing, her skin was sweaty and blue and she had a fever of 40°C. Ibsen instructed a surgeon to insert a tube into her windpipe under local anesthesia, and then attach a rubber balloon to the end of it.

When Dr. Ibsen started pumping air into Vivi's lungs, it did not look well at first. It seemed the girl would die at any minute. Doctors observing the procedure started leaving the room. But then Dr. Ibsen injected the girl with a medicine that prevented her from breathing on her own, which made distributing air into her lungs a lot easier. After this modification, the girl's temperature and blood pressure dropped and her skin went back to its normal color. Vivi E. became the first patient whose life was saved during the polio epidemic by means of a ground-breaking method that still constitutes the basis of intensive care.

To convince the sceptics, Dr. Ibsen then put the patient into an iron lung, which swiftly convinced everyone present that the machine's negative pressure system is inadequate. It simply doesn't supply enough oxygen to afflicted organs. Dr. Ibsen thus demonstrated that efficient assisted respiration requires a catheter, the suspension of autonomous breathing and the steady pumping of air into the lungs with a balloon.

When it became clear that this simple method could be used to save the lives of many children struck by the epidemics, the hospital started recruiting medical students and later even dental students to administer the treatment. A direct result of this radical method of dealing with the deadly disease was the drastic drop in the mortality rate mentioned earlier, falling from 92% of the children at Blegdams hospital requiring the oxygen treatment to just 25%.

The news of the method quickly spread all over the world. The hospital in Denmark soon received a visit from Oxford neurologist Ritchie Russell MD, who wrote a classic monograph giving instructions on how to set up the polio treatment. He tested the method in his own hospital on a sixteen-year-old schoolgirl suffering not from polio but a severe nervous infection that paralyzed everything but her eyes. Challenging the medical doctrine of the time, he instructed a surgeon to insert a catheter into her windpipe and start artificial respiration. After a few weeks, the infection subsided, her strength was restored and she eventually recovered. She later became a nurse and a mother of four.

How nerve poison saves lives

An important detail in the procedure that Dr. Ibsen transferred from operating theatres to intensive care wards was the use of Curare, a substance which paralyses the respiratory muscles and thus prevents patients' bodies from resisting artificial respiration.

Curare is better known as a nerve poison used by South American Indians in hunting. It was first deployed for medical purposes in an operation in Montreal in 1942. A few years later an anesthesiologist in Liverpool discovered that the nerve poison greatly reduces the dose of anesthetic required during surgery.

Unfortunately the method of assisted respiration by means of catheter wasn't used widely enough to treat other serious conditions. It would have greatly benefitted patients recovering from difficult operations, but surgeons were very sceptical about using it. They were worried that the patient's helpless state after surgery would make it seem like they hadn't done a good enough job, so they refused to insert catheters into their patients' throats. But after a few years it became clear that an insufficient supply of oxygen due to respiratory dysfunction was a major cause of post-surgical complications.

In 1953 in Copenhagen, the anesthesiologist Bjørn Ibsen MD founded the world's first permanent intensive care unit The first device used to supply the patient with oxygen artificially is still considered one of the key machines in intensive therapy today, enabling critically ill patients to get through the worst phases of their disease while their body fights for survival, and to regain all their normal body functions. The importance of oxygen to human physiology has been known for at least two hundred years, but its crucial role in preserving the lives of the critically ill became conclusively apparent during the polio epidemic of 1952 in Copenhagen.

Today polio is endemic only in Pakistan, Afghanistan and Nigeria. All other countries have successfully managed to eradicate it with methodical vaccination programs. These programs were prohibited in the three countries mentioned above, where certain Islamic religious groups declared that the medical staff administering vaccines were enemies of the state and even initiated attacks that ended in massacres.

SOCIETY

Three stages of revolt

After her day's work at a department store, forty-two-year-old seamstress Rosa Parks from Montgomery Alabama took her bus home. She got on a city bus as usual, paid the driver ten cents for a ticket and looked around for an empty seat to rest her tired feet. It was the December 1, 1955, a little past six in the evening, and nobody could guess that the events that were about to unfold would eventually reshape the United States of America.

According to regulations of the day, the first four rows on the bus were reserved for white passengers, with members of the black population forced to cram into the back. Parks took a seat right behind the "white zone", where people of both races were allowed to sit.

With the next few stops, more people got on and eventually the front part of the bus became crowded too, leaving no empty seats for white passengers. When the driver, James B. Blake, noticed that a white man was standing, he yelled at the black passengers to move further to the back and give more space to the white people in the first four rows. But nobody moved. The enraged driver stopped the vehicle and went to take care of the situation himself. Following his orders, some of the black passengers stood up and gave up their seats for the white man who was standing, but Rosa Parks maintained she was not sitting in the "white zone" and refused to leave her seat for one white man, since three seats were now already empty.

But the driver wouldn't let it go. He commanded Rosa to give up her seat, or else he would call the police. Since she kept resisting and demanding an explanation, the police came, arrested her and took her into custody.

The birth of a mass movement

Little did the bus passengers know that the incident would mobilize a mass movement that would soon spread throughout the US. In reaction to Parks' arrest, African Americans in Montgomery began boycotting city buses, causing the municipal transport company to sink into serious financial trouble, while a charismatic young pastor called Martin Luther King, who had moved to Montgomery only a year earlier, passed on the spark of revolt to other cities, eventually reaching the capital.

Interestingly, upon arriving in Montgomery, Martin Luther King observed that the African American community generally accepted racist rules and legislation, including violent treatment, without putting up much resistance. So what was the turning point that made them stand up against the obvious injustices? What was the key difference between Parks' rebellious act, which triggered a mass movement, and courageous protests by other individuals before her that had left no mark? Parks was not the first person to oppose the racist regulation governing public transport and other basic civic areas. There were many similar stories but none of them inspired a mass uprising.

Spontaneity, emotion, solidarity

According to Charles Duhigg, the author of *The Power of Habit* (Random House, 2012), the fact that Rosa Parks was a widely known and respected citizen played an important role in people's reactions to her arrest. The news spread fast and instigated a feeling of unease throughout a wide circle of her acquaintances.

Mass social movements never result from a sudden realization by the majority that some things need to change; they are almost always triggered by a concrete event that sets adrenalin flowing. It is important for protesters to channel their pre-existing motivation. It is crucial that their reaction be spontaneous and emotionally driven, since this is the only inner drive strong enough to sustain a rebellion.

In the next stage, this authentic emotion needs to be transformed into a common movement supported mostly by interpersonal bonds between protesters. In this stage, supporters of the movement don't insist on their demands being met solely for the sake of the event that provoked their rebellion but, above all, out of solidarity with each other. For a revolt to be successful, a common opinion needs to be formed that the protests involve an act of honor, one worth fighting for. If there is no peer pressure that forces the protesters to stand their ground, the movement is bound to die out.

The third stage of revolt enables protesters to surpass their local environment, and calls for new patterns in everyday conduct that spread throughout society as something more or less taken for granted. It is only at this stage that an uprising causes permanent change in society.

Changing mass habits

In his book, Duhigg describes three stages of change in social habits that are all equally important. First: a definitive event that is emotionally charged and gets people to their feet. Second: the growth of a conviction that opposing the protest is dishonorable, which motivates people to keep on protesting even when their initial motivation wears off. Third: the formation of new rules and habits that differ from the previous laws and conventions and are adopted by the majority of the population.

Society changes only when we change our habits or the unconscious patterns that we follow when we're not paying attention – when we are moving on "autopilot". Even if the reasoning and the ideas behind the change are completely clear, a society won't change until people internalize new patterns of behavior.

History teaches us that the ideas of Voltaire or Rousseau circulated for ten years among the intellectuals of the time, but only when the common people started to discuss issues such as universal human rights and the corruption of the ruling monarchies did the absolutist regimes across Europe start to collapse.

Similarly, a handful of courageous suffragettes agitated for women's rights for decades, yet the laws weren't altered until the equality of women started to feature prominently during family dinner conversations.

A long way towards change

The Montgomery bus boycott lasted for nearly a year. It included a number of violent interventions by the police and mass arrests, but people seemed to lose their fear completely. Being part of the revolt was more important than fearing the blows and intimidation mustered against it. The protesters even went so far as to give themselves up voluntarily to the police to show how meaningless these fear-driven acts of authority really are.

Even though the US Senate voted on June 5, 1956 that racial segregation on public transport was unconstitutional, the city of Montgomery appealed its decision and it was only on December 17 that the US Supreme Court ruled conclusively against the segregation. Three days later, the municipal authorities received an order to ban racial discrimination on city buses. The next morning at 5.55 a.m., Martin Luther King and his colleagues mounted a bus for the first time after a year, took their seats at the front end of the bus and went for a ride through the streets of Montgomery.

The success of the Montgomery uprising encouraged a mass civil rights movement that continued to change America through peaceful protests. The supporters of the movement were still met with violence and it took almost a decade until President Lyndon

Johnson finally signed the Civil Rights Act, which prohibited any form of discrimination against ethnic minorities and women, in 1964. In his speech on this historic occasion, he proclaimed that the civil rights pioneers were the founding fathers of modern America.

The wisdom of crowds

In the fall of 1906, Sir Francis Galton, an all-round British scientist and relative of Charles Darwin, visited a country fair. Despite his venerable age of eighty-five, as a statistics enthusiast he was attracted to a guessing-game competition which invited fair-goers to bet on the weight of a soon-to-be slaughtered ox. As many as eight hundred visitors placed their bets by buying a specially numbered and stamped form on which they gave their personal details and an estimate of the ox's weight. The person with the nearest guess won first prize.

As Galton later reported in *Nature* magazine, the participants included professional livestock breeders and butchers as well as random fair-goers with no previous experience in estimating an animal's weight:

> The average competitor was probably as well fitted for making a just estimate of the dressed weight of the ox, as an average voter is of judging the merits of most political issues on which he votes, and the variety among the voters to judge justly was probably much the same in either case.

Galton was primarily interested in how good an answer an "average voter" can provide when answering a difficult question. Galton himself was a keen proponent of the superiority of the intellectual elite and was convinced that the masses are incapable of great achievements. When the prizes were distributed after the competition, Galton borrowed the voting sheets and compared them for statistical analysis. He eliminated thirteen incomplete sheets and used the remaining information to calculate the average estimated weight the crowd at the fair had given for the ox.

He expected the average estimation to be far off the actual weight of the ox, since it was mostly contributed by people who weren't experts in the field. But he was wrong. His analysis showed that the crowd's average estimate was 1197 pounds, incredibly close to the animal's actual weight of 1198 pounds. Galton, a fan of eugenics among other things, was surprised by the finding: "This result is, I think, more creditable to the trustworthiness of a democratic judgment than might have been expected."

The stock market's reaction to the space shuttle Challenger disaster

On January 28 1986, twenty-two minutes before noon local time, the space shuttle Challenger took off from Cape Canaveral, Florida. A minute later it exploded more than fifteen kilometers above the earth. Since the launch was broadcast live, the news of the tragic death of NASA's astronauts spread like lightening. Less than ten minutes after the disaster the stock of four major companies involved in building the shuttle began to drop.

Interestingly, within the next few hours stocks in Morton Thiokol, the company that built the shuttle's solid rocket motors, dropped by nearly 12%, whereas the stocks of the other three companies swayed at first but eventually settled at 3% below their initial value.

In a celebrated analysis of the stock market's reaction to the disaster, Michael T. Maloney and J. Harold Mulherin showed that within hours of the accident the market already "knew" who was responsible, even though it took a special commission of experts several months to establish which parts of the shuttle had actually failed. The research commission was specially assigned to the case by the US president and included a number of esteemed astronauts, pilots and constructors, among them the recipient of the Nobel Prize for physics, Richard Feynman. At a press conference six months after the accident Feynman presented the commission's final report. He demonstrated a small experiment in front of the cameras to show how a low temperature caused a leak in one of the solid rocket motors.

Interestingly, the two stock market researchers couldn't find out how investors knew so quickly which company was responsible. At first they presumed that company employees had quickly realized their mistake and tried selling their stocks, but no such transaction was discovered. Moreover, Thiokol's stocks also weren't even sold by employees at rival firms, who had expert knowledge as to possible causes of the accident and weren't bound by loyalty. It seems instead that investors intuitively refused to buy stock from the company at fault.

In his book *The Wisdom of Crowds* (Doubleday, 2004) James Surowiecki discusses both the stock market's response to the Challenger disaster and the estimated weight of an ox at an English country fair as two examples of "crowd wisdom". Surowiecki's theory explains the mechanisms behind the crowd's extraordinary intelligence in certain situations, their achievements greatly surpassing the individual capabilities of its most competent members. Even if no member is capable of great achievements on their own, the crowd is able to reach exceptional results by means of a collective decision-making process.

Another example mentioned in the book is an unusual sight, observed in 1921 in the jungle in Guyana by the American natural scientist William Beebe. He noticed a group of ants moving in a large circle more than three hundred meters in diameter. Each ant took two hours and a half to walk the entire circumference. They were taking no breaks, rushing as if they were going somewhere, even though they never reached their destination. In a few days all the ants died of exhaustion.

Conditions for collective intelligence

"The circle of death" described by Beebe is now explained by biologists as a phenomenon in which a group of ants separated from the main foraging party begins to follow each other's pheromone track, forming a continuously rotating circle. Ants developed this strategy through evolution, and in most cases it results in a happy ending, since it enables them to find their way back to the colony. But when the leaders of the caravan start to follow their own track, the strategy proves disastrous. An individual ant is completely dependent on the direction taken by the one in front, and that is why all the ants soon start running in a circle until they eventually die of exhaustion.

Surowiecki uses this unusual behavior of ants and their continuously rotating circle as an analogy for a group of people who depend solely on one opinion of their surroundings, with their subsequent actions often resulting in tragic consequences. A crowd of humans is always in danger of endorsing and spreading false beliefs and ideas are when they are strongly influenced by group mentality. When faulty propositions are made by random individuals, the crowd has no trouble eliminating them. But if the majority of the group is in the wrong, and if it doesn't function as a group of independent individuals anymore but rather as a homogenous whole, it becomes unable to correct mistakes in the process of reaching consensus.

There is an important condition that needs to be met in order to make a crowd intelligent: the individuals that form it need to think and function as independently as possible and rely predominately on their own knowledge and abilities. This recommendation was certainly in place at the weight estimation competition, since its participants didn't share their opinions among themselves. It also probably holds true of the stock market reaction to the Challenger disaster.

In addition to a wide array of independent thinkers, it is indispensable for a group to have an efficient system of synthesis in order to transform individual opinions into a collective stance or decision. Google for instance is very successful at arranging web pages according to their relevance, primarily because it relies on independent evaluations made by an enormous number of internet users. Anyone who uses their browser or posts anything on the net is contributing to the quality of the suggested hits. Google's PageRank technology is responsible for assessing the relevance of a web page solely by analyzing all the connections between it and other web pages. The company states that:

> PageRank relies on the uniquely democratic nature of the web by using its vast link structure as an indicator of an individual page's value. In essence, Google interprets a link from page A to page B as a vote, by page A, for page B. But, Google looks at more than the sheer volume of votes, or links a page receives; it also analyzes the page that casts the vote. Votes cast by pages that are themselves "important" weigh more heavily and help to make other pages "important."

It is extremely important when establishing quality mechanisms in collective decision-making to keep people from paying too much attention to other people's opinions and to encourage them to rely chiefly on their own experience, information and reasoning. The opinions should be kept independent and shouldn't be formed purely in opposition to other views. It is not the diversity of opinions that counts, it is the diversity of procedures and ways they are formed that is crucial to society.

What money can't buy

For a long time, the citizens of Switzerland struggled to find an appropriate place to dispose of their radioactive waste. After taking all the circumstances into consideration, the experts suggested a tiny village in the middle of the country as the most suitable location. Shortly before a local referendum was held, villagers participated in a survey that showed a minimum majority of 51% was willing to endorse the building of a waste disposal plant. But an additional question in the survey provoked a most surprising reaction. The villagers were asked whether they would be more inclined to support the plan if they received government compensation for their trouble. Even though this offer would commonly be perceived as more acceptable, it was accepted by only 25% of those polled.

Introduction of payment changes the rules of the game

Similar, seemingly irrational behavior was observed by kindergarten employees in Israel. In trying to get parents to come and pick their children up from kindergarten before the official closing hours, they had introduced monetary fines for latecomers. But soon after the new rules came into force, the management was surprised to find that overtime fees caused the unwanted behavior to increase rather than diminish. Now even more people than before were late collecting their kids.

When staff discussed the issue with the parents, the reason for these unexpected results soon became obvious. Before the overtime fees were introduced, parents who came in late felt guilty about breaking the kindergarten's code of conduct and at least tried not to be even later than they usually were. When the fine was introduced, parents perceived it as an extra payment which entitled them to be late once in a while. Paying the fine made them feel less guilty about showing up late for their infants and toddlers, and so they made less of an effort to avoid it.

To get a better insight into such anomalies, Israeli researchers conducted an experiment by ordering a group of students to collect donations for a charity fund. The results surprisingly showed that the most successful students were those who received no monetary compensation for their work. A group of volunteers collected as much as 55% more donations than their colleagues who were awarded 1% of the funds they collected. The group that received 10% of collected funds achieved better results but still had less success than the volunteers.

All three of the cases above are discussed by Harvard professor of ethics Michael J. Sandel in his book *What Money Can't Buy: The Moral Limits of Markets* (Farrar, Straus and Giroux, 2012). The author stresses that market orientation is by no means neutral in value. When a certain set of goods or services reaches a value that surpasses its market price, this valuation influences our attitude towards it. We often don't realize how a mere readiness to trade with something completely changes its value and how we perceive it.

After the Cold War, when a market-based method of producing and distributing goods was globally accepted as the most successful and effective economic model, market-based thinking and values began seeping into other areas of society.

Market economy and market society

In order to shed light on several decades of social changes, Sandel introduces a terminological distinction between a market economy and market society. A market economy is a very efficient tool for organizing the production and distribution of material goods and services, while the term market society applies to the market's way of assigning value in all areas of society.

The transformation from market economy to market society takes place when we start to apply market values to areas of life that cannot be defined in such terms. This imprudent propagation of market-based thinking, commonly referred to as neoliberal ideology, consists of the idea that market-value attribution is the most fair and effective, regardless of the area of society to which it is applied.

Sandel finds that in a society where almost everything is for sale, two big problems occur. Social inequality increases, placing the rich in an over-privileged position, and a damaging effect of trading is exposed. As seen from the examples above, trading strongly influences the way we perceive and value a certain area of society when market-based valuation is introduced.

Another consequence of increasing the market-based approach is that the ways of managing social relations become less varied, making them less meaningful and less legitimate. The most illustrative and common example of a "non-market-based" management of social relations is the everyday custom of standing in line. The ethic of queuing is based on the principle of "first come first served", a fundamentally egalitarian maxim that presumes all those in the line are equal in their rights and differ only by the place they currently occupy.

Market orientation is of course egalitarian on its own terms, since it requires you to have money in order to do business, regardless of who you are. Everyone who is able to pay is, according to market ethics, equal. But unlike standing in line or a lottery draw, anyone who lacks the necessary means to trade is cast out of the market.

The idea that the market system is fair and effective relies on the assumption that business is conducted only when both sides end up better off after the trade, or at least maintain their status quo. The image of society as one big market mechanism supposedly depicts all of its members consciously or unconsciously calculating their losses and gains with every decision they make, be it a monetary exchange or in managing interpersonal relations. Presumably, other key factors in market functioning are that all who take part in a trade are entirely free and increasingly well informed.

What we shouldn't trade

Sandel stresses that the pillar of contemporary society is based on a restriction in trading certain goods. For instance, nobody can sell their right to vote. Nor can you hire someone to serve your prison sentence for you. We could say that the core of trade restriction rests on the idea of the true equality of all people, which is the basis of most modern social institutions, including the judicial system, political decision-making, science, education, social rights, health care, etc.

If we take science as an example, it is absolutely imperative for it to function outside market valorization. The sole thing that counts in science is the argument, regardless of who produces it and how much they would be prepared to pay for it being accepted by the scientific community. Science is, at least in principle, an example of a modern societal institution, based on complete egalitarianism and a total rejection of market principles, since all scientific knowledge is by definition public and available to everyone. The valorization of ideas in science is based on the principle that theories may be refuted by anyone who provides a solid argument and this is precisely why they should be publicly available. Introducing market principles to science itself is therefore by definition impossible.

But restricting the growth of a market society doesn't only seek to protect the basic institutions of contemporary society; it also addresses the corruptive influence of trade in areas where the introduction of trading principles is capable of causing a change in our values and our attitude towards a specific issue.

Sandel spells out the damaging effects which ensue when mechanisms allowing trade in greenhouse gas emissions are introduced. Following the analogy of latecomers in kindergartens, the possibility of paying a fine eliminates the feeling of guilt from socially undesirable behavior. Allowing companies to 'trade' their toxic emissions in exchange for a payment, service or some other penalty transforms something previously perceived as morally unacceptable into a legitimate and marketable commodity. To illustrate this, imagine as an extreme example a situation in which anyone could dump their trash wherever and whenever they pleased, provided that they pay a person living in poverty in the Himalayas to pick up a corresponding amount of trash near their own home. Would that change your view of outsourcing?

The Stanford prison experiment and the Lucifer effect

On a hot Sunday afternoon in 1971, an unusual experiment took place in a basement in Stanford University. Philip Zimbardo, a psychology professor, recruited young men and paid them fifteen dollars a day to participate in some scientific research. More than seventy candidates responded to his advertisement, since it was the vacation, and the experiment seemed like an easy way to earn some extra cash.

During their preparations for what became one of the most notorious experiments in the history of social psychology, Zimbardo's team screened all the candidates in great detail. Twenty-four respondents were chosen, all of whom had "ordinary" or "normal" backgrounds, no criminal records, broadly middle class origins and nothing that would make them stand out from average citizens. Among the accepted candidates, nine were randomly assigned the role of prisoners (by a flip of a coin), and the other fifteen were given the role of prison guards. The experiment consisted of a two-week-long simulation of prison life in the basement of the Stanford psychology building.

Arrests were made by real police

The researchers were primarily interested in what happens if good people are put in bad situations. They planned to assess the influence of the prison environment on the relationships that formed within it and compare what happened to how they expected the inherent personal traits of the guards and prisoners involved would lead them to behave normally. The results of the experiment, which still brings up unpleasant memories for everyone involved, were horrifying. After only a few days, the conditions in the simulated prison became unbearable and the experiment had to be terminated after only six days, less than half-way through its projected timespan.

To make the situation as true to life as possible, Zimbardo arranged with the head of the local police to have real officers arrest the "prisoners" at their homes on a Sunday morning, before they even knew which role they would play in the experiment. Their astounded neighbors watched them being handcuffed and taken to custody. At the police station, they were treated like real prisoners; the officers took their fingerprints and brought them to the university's mock prison blindfolded. The prisoners were met by the first shift of prison guards, who were divided into three shifts of five. This way each of them spent no more than 8 hours a day on the job, a length of time consistent with the working hours of real-life prison guards. The experiment leader, Philip Zimbardo, took on the role of prison superintendent.

Upon arrival, the prisoners were systematically dehumanized. They were stripped of all personal belongings and assigned numbers and identical uniforms. The guards wore official uniforms and prisoners had to address them as "officer" or "sir". They also wore reflective glasses to prevent any direct eye contact with the prisoners.

They took to their roles like ducks to water

The first prisoner to break down fell into severe depression after only thirty-six hours in prison. He suffered ongoing attacks of uncontrollable rage and weeping. After a long discussion, the researchers decided to send him home. In the course of the following three days, three other prisoners had to be released due to similarly severe cases of anxiety; one further detainee had to leave the experiment because he developed a psychosomatic rash all over his body.

Even though both the guards and the prisoners were ordinary young men, who could leave the experiment at any time, it took them only a few days to accept their roles so completely that they forgot the entire thing was a mere simulation. The duties of the guards were at the beginning very simple, since their only obligation was to maintain order. But on the second day, the prisoners refused to carry out the tasks they had been ordered to perform. They barricaded themselves inside their cells and started verbally abusing the guards. That was when the trouble started, with unexpected consequences.

The guards strived to retain their authority by any means possible and make sure the prisoners abided by the rules. To this end, they put the presumed leader of the rebellion into solitary confinement, while the other prisoners had to do push-ups, jump up and down, scream and verbally abuse one another. Researchers filmed everything that was going on in the mock prison (allegedly without the knowledge of the young men involved), in order to analyze the situations and reactions of both parties. Some of the footage from the Stanford basement in August 1971 is available online at *www.prisonexp.org*, along with a detailed description of the experiment.

The experiment becomes unbearable

Interestingly, Zimbardo himself entered so strongly into the role of superintendent that he was more concerned about the trouble-making inmates and an attempted break-out than in taking notes and conducting the experiment. When one of the guards overheard prisoners talking about an escape attempt, they became so worried that Zimbardo, still putting his position as superintendent first, actually went to a real police station and asked the officers to lend them a real police detention unit for his most rebellious prisoners. Fortunately, the police refused to cooperate, but Zimbardo's act proves how far he had gone in getting into his character.

He was brought back to reality by his girlfriend, Christina Maslach, who had recently finished her doctoral studies and was helping him with the experiment. She was the only observer to suggest that the people involved had gone completely overboard and that the experiment was no longer about science but victimizing a group of innocent young men. After six days she gave Zimbardo an ultimatum: either he ended the experiment or she would leave him. Only then did Zimbardo realize how serious the situation had become. He terminated the experiment on the August 20, 1971, after six days and six nights.

Lucifer's metamorphosis

The Stanford prison experiment showed that insane situations triggered insane behaviour in otherwise completely normal individuals. Philip Zimbardo has recently published a book entitled *The Lucifer Effect: Understanding How Good People Turn Evil* (Random House, 2007). Thirty-five years after the experiment, he describes and analyses in detail everything that went on and compares the results with corresponding phenomena in contemporary society. He is mainly interested in how average individuals turn into vicious criminals, ready to harm other people deliberately, induce physical pain or even cause death. This transformation from good to evil inspired the book's title. The Lucifer effect, named after the angel who turned on God and became Satan, describes the ease of passage from good to evil.

In his book, Zimbardo describes in detail the abuse of Iraqi prisoners by US soldiers in the Bagdad prison of Abu Ghraib. He testified as an expert defense witness at the trial of one of the soldiers, so could accurately describe the events that took place. It was stunning how a loving family man became an incarnation of pure evil under vicious circumstances. Zimbardo is primarily interested in what are the key environmental elements that trigger the Lucifer effect, the transformation of good into evil.

There are no simple answers. On his website *www.lucifereffect.com*, Zimbardo describes a few methods for examining our reactions in different situations in order to resist the Lucifer effect.

The events that took place that summer in Palo Alto were dramatized in a German movie, *Das Experiment* (2001), and a Hollywood version is on the way. The most preposterous thing about this experiment is that it was actually approved by Stanford's ethics committee, whose main concern was about the University basement's fire safety.

Why is one environment more creative than another?

During World War II, a wooden two-story building was constructed on the campus of MIT to house scientists and engineers arriving in Boston from all over the world. The shared aim of these immigrants was developing new military technologies for the Allies. "Building 20" was intended as a temporary structure, and no one was particularly worried about its appearance. It was described as a plain and ugly shack, and legend has it that the blueprints for it were roughly drawn up in no more than a day. Despite its makeshift nature, the building survived until 1998. By this time it was known as "the magical incubator", and celebrated as one of the most inspirational edifices ever built.

Immediately after the war, with demolition already imminent for Building 20, it turned out that extra facilities were needed to accommodate the large inflow of new students. So the building was assigned to researchers and professors who couldn't find a place for their offices and laboratories elsewhere. Building 20 became a melting pot of scientific researchers from many different fields who hardly knew each other.

The magical incubator

This interdisciplinary mixture of experts sharing one roof may seem like a recipe for disaster. Yet it proved quite the opposite, since the building was a place of constant discovery. Among other advantages, the temporary wooden structure was easy to change and adjust. When Jerrold R. Zacharias set out to develop his atomic clock, he needed a laboratory that could house a large pole, so he simply took down the ceiling between two floors. Those who needed bigger rooms took down the walls and adjusted the space according to their needs.

In 1956 Amar Bose, a music-loving inhabitant of Building 20, decided to enjoy a little divertissement while writing his doctorate. Upon buying a record player for his office he was disappointed to find that even the best speakers on the market couldn't produce a sound to compete with live music. He became interested in the technologies used to construct speakers and visited the acoustic lab, which was conveniently located in the same building. He soon spent more time in the lab than at his own desk. The result of Amar's interest was the Bose Sound Company, still the leading producer of audio systems, and a personal fortune for him amounting to billions of dollars.

But natural and technical scientists were not the only users of Building 20. "Two of the most miserable holes" in the building, as their tenants described them, were occupied by the founder of the department of linguistics, Morris Halle, and a young doctoral student, Noam Chomsky. The latter caused a real revolution in linguistics and is considered one of the most cited living scientists.

The power of coincidental encounter and meeting new faces

Many have long wondered what makes one environment more "creative" than another. One hypothesis stresses the importance of coincidental encounters between people running into each other on a daily basis and exchanging a few words. One of those who followed this wisdom was Steve Jobs, who applied it when establishing a base for Pixar Animation Studios, the birthplace of many top feature films that redefined computer animation. To increase the rate of spontaneous encounters between employees, Jobs arranged offices in the complex around a glass atrium, a central space where impromptu meetings could be held and where Jobs eventually installed the employees' pigeonholes, a coffee bar and all services that were frequently used by studio staff. It is said he even closed off the other toilets in the building to make employees use those in the atrium, but the idea didn't last very long.

Testing the hypothesis that creativity increases with direct contact involved a purely quantitative method. Isaac Kohane studied the distance between the places where authors of successful papers had carried out their research. After analysing 35,000 scientific publications and identifying where the research was done, Kohane found a clear connection between the vicinity of the authors and the frequency with which they were quoted. The best results were consistently attained by teams of researchers working within tens of meters from each other.

In addition to proximity and chance meetings, to sustain a creative environment it is important to provide a constant inflow of new faces, new experiences and new knowledge. Sociologist Brian Uzzi wanted to find a magical formula for putting together a creative team with the maximum chance of success. He started by analyzing the history of Broadway musicals, where he gathered reliable information about the success rate of each show as well as its full cast and crew.

After studying the production teams of nearly five hundred different Broadway shows staged between 1945 and 1989, Uzzi could empirically prove that a team's diversity is the key to its success. He introduced the Q factor, measuring the extent to which the people working on a show already knew each other. If the team members were barely acquainted, their Q was low, whereas if they had worked together before, this resulted in a higher level of Q.

The results showed a significant correlation between the Q factor and the show's subsequent success. On the Q scale between 1 and 5, a Q lower than 1.7 and higher than 3.2 meant the collaboration was not going to be particularly fruitful. The golden mean proved to lie between 2.4 and 2.6, indicating that the best results are achieved by a proven team of old players when joined by some new faces.

Teamwork is today a common practice in science. That is why Uzzi, together with management professor Benjamin F. Jones, continued his analysis and studied twenty million academic papers along with two million patents published during the last fifty years, in order to identify a possible correlation between the size of a research group and its efficiency. The results, published in 2007 by *Science* magazine, showed that the size of research groups increases every decade by approximately 20%. It turned out that papers of shared authorship are cited twice more often than those published by a single author. The difference is even greater with the sought-after articles that get quoted a hundred times or more. Such papers are six times more likely to have been written by a team than an individual researcher.

Why brainstorming doesn't really work

There is little doubt, then, that interaction and proximity are highly conducive to fruitful research. But the evidence indicates that simply putting people and their ideas together in a relaxed environment is not an answer in itself. A need still exists for individual space and then critical input, as the story of one of the best known forms of intellectual teamwork strongly suggests.

When discussing creative collaboration, the first idea that comes to mind is the method of searching for ideas described soon after World War II by Alex Faickney Osborn, in his bestselling work *Applied Imagination*. A New York advertising executive, Osborn named the technique "brainstorming" and defined its main principle as "ideative efficacy". This could be attained by, among other things, successfully withholding criticism that might have a negative effect on other people involved.

The method soon became widely popular with businesses that required a constant supply of fresh ideas. However, experts researching the field of creative processes remained unconvinced. When at the end of 1950s they tested the technique on a group of students, it turned out the teams that followed the brainstorming principles did much worse than individual students working on the same task. On average, individual students came up with twice as many ideas as their brainstorming colleagues. Judges also gave the individual thinkers' ideas a higher grade, on the whole, than those of the brainstormers. According to Jonah Lehrer's article "Groupthink – The Brainstorming Myth" (*The New Yorker*), several other studies reached the conclusion that brainstorming was harmful or at least inefficient. Such findings were consistently ignored by the creative industries.

Additional research showed that the most problematic principle of the brainstorming technique was precisely the one that requires participants to withhold their criticism. New findings show that it is important to face criticism in the creative phase of the working process, because it enables us to see the matter at hand from different viewpoints. Constructive criticism is not a foe but a friend to creativity.

The equation that almost destroyed capitalism

A rule of capitalist economics states that one should be able to take risks in order to gain profit. When making business decisions, we always need to take into account the possibility of losing our investments. In principle, the more chances you take the bigger your potential profits. Or, alternatively, be ready to lose on a large scale if the business doesn't turn out as planned.

It seemed almost impossible to get around this defining trait of capitalism in order to ensure oneself against potential loss, without breaking any laws. But a few decades ago, economists managed to come up with a magic mathematical formula that was able to do just that. By the use of one simple equation, they could control stock market risk and guarantee long-term profit. The authors of the formula achieved fame, made a fortune and even received the Nobel Prize for economics in 1997; yet despite the odds of successfully controlling the risks, not everything went according to plan.

Trading with the future

Metaphorically speaking, this "magic formula" enables us to adjust our bets at a horse race, before the race is finished. Clearly, we would all be more willing to raise our bet on the horse that is leading the race than on one that is trailing behind. But when there is a new development on the field, the odds on certain bets, or trading options, change accordingly.

The equation uses the information about what is going on in the race to calculate which investments would pay off and which wouldn't. Stockbrokers use it to assess current risks and to set a corresponding rate for each option contract, based on current information.

Field crops are a typical example of an options market. A farmer sells his wheat crop to a food-processing factory before he actually sows the seed. The factory and the farmer set the price, the quantity and the date of delivery in advance, before the wheat is harvested. In the meantime, the factory is allowed to sell its option contract to another factory, one that wasn't able to seal a deal in time and is now in need of wheat supplies. This other factory is prepared to pay the first factory extra money for the contract, which will enable it to buy wheat at a set rate during harvest time.

This means that a company making a number of such contracts without really needing all the wheat they were buying, would make a lot of money in case of drought or another natural disaster increasing the market price of wheat. The contract binds the two parties to conduct their business at a set rate regardless of the fluctuation in the market price of wheat. If the rate increases, the profits grow too, but so do the losses if wheat loses market value through overproduction.

Option contracts are a combination of insurance and a market-based approach. The farmer and the factory are both insured with a contract that guarantees them the payment or supply they are due. It enables them to stay safe by avoiding any risk regarding the final market price, and to sell their contracts at any time if this constitutes a good business opportunity.

The Black-Scholes equation

The magic formula, designed to control risk more efficiently, was first published in 1973 by Fischer Black and Myron Scholes, and became known as the Black-Scholes equation. It is a partial differential equation, and the higher mathematics it involves is of a type usually used only by technically educated scientists and engineers. As a result of this formula investment banks started recruiting postdoctoral students trained in solving the most complex mathematical problems.

For the Black-Scholes equation to bear useful results, it is very important to know exactly how likely it is that the prices of goods to which we want to apply the mathematical model are going to change. We can't know exactly what a future price is going to be, but to a certain extent we can predict whether it is about to rise or drop. These changes, and the rate of overall change, can be quite accurately predicted in short-term stable situations, although unanticipated events (such as natural disasters or terrorist attacks) can change the picture drastically.

It may seem like a contradiction, but coincidence in itself isn't a problem. The key issue with the Black-Scholes equation is the fact that it works well only if predictions are based on correct information. It works well under stable conditions; that is, when the stock market developments are influenced by a number of small, unrelated events which mathematical modelling can account for easily. Up to a point, statistics has little problem describing coincidences that are totally unrelated.

The equation is therefore applicable when conditions are relatively calm, and when events are distributed by "predictive coincidence" or when the rate of fluctuation in the stock market stays within the boundaries of reasonable expectation. It is when unexpected events influence such "predictive coincidences" that some serious issues can occur.

When Nobel Prize winners enter the stock market

One of the first scientifically endorsed investment funds was called Long-Term Capital Management (LTCM). Two 1997 Nobel Prize-winners in economics were actively involved in creating it; Myron Scholes and Robert C. Merton. They were awarded the prize precisely for their contributions to developing and improving the equation used to calculate risk in stock-market transactions. The equation was used daily by the investment fund, while the name and reputation of Scholes and Martin attracted numerous investors and assured safe transactions.

The fund boasted large profits at first, attracting ever more investors, but unfortunately the "magic formula" on which all the business decisions were based eventually failed to predict or consider the risk of unexpected or, as they are known in investment circles, "black swan" events.

Between 1994 and 1998 the LTCM fund made a good return for its investors, but suddenly went into complete freefall in a matter of months. The stock-market crash of 1998 revealed the flaws in the magic stock-market formula, and it turned out not to be as magnificent as everyone had thought. One of the causes for the crash was the Russian financial crisis which occurred when a number of abrupt measures were taken by the Yeltsin administration. The Kremlin announced that Russia would not pay off its debts until the situation was resolved. This triggered a domino effect destabilizing the stock market, which wasn't something the risk management mechanism could have included in its calculations of predictability.

The Russian financial crisis of 1998 doesn't exactly constitute an extreme example of a black swan, but it nevertheless brought the investment fund to its knees. Blinded by the "magic formula", the fund lost most of its assets in only a few months. As the LTCM sustained losses of nearly five billion dollars, the U.S. Central Bank organized a consortium of 14 private banks to bail it out. They were certain at the time that the fund was simply too big for them to let it go bankrupt, and the consequences for the banking system couldn't have been foreseen.

The Black-Scholes equation is still used today and offers an efficient tool for risk control in most circumstances, except of course for true black swan events. In those cases, the tool becomes unreliable and can cause a many problems. Modelling the stock market mathematically is not the same as modelling atomic or planetary movement. Isaac Newton himself realized this on losing his savings to one of history's first stock market crashes. "I can calculate the motion of heavenly bodies," he declared, "but not the madness of people."

Autocorrect

Shortly after World War II, while carrying out maintenance on one of the world's first computers in Bell's laboratories, the mathematician Richard Hamming was faced with issues that hardly ever arise today. Since a single computer was used by more than one scientist, he needed to reprogram the computer and reintroduce the processing data every time his team wanted to get on with their work. Today, this is done in a flash and usually without any complications, but in the era of computer science's pioneers, these tasks were still a complicated matter.

Algorithms for detecting and correcting errors

The problems were caused primarily by frequent errors in the computer's data processing, causing major complications and stress for researchers. If there was a single error in the reading, all the calculations proved useless. Moreover, it was very likely that the computer would stop responding in the middle of running a program and thus need to be reset.

Next to calculations and algorithms, storing and transmitting information are the main tasks computers perform. This requires extreme levels of accuracy. The data needs to be exact, with 99.99999% reliability, or problems will be quick to occur. Achieving absolute accuracy was an important issue for the first computer-users after World War II and needed resolving as soon as possible.

In 1947, in an attempt to avoid the recurring troubles in data reading, Richard Hamming discovered an algorithm that was capable of detecting and correcting errors to a limited extent. Even though the news of his discovery was published only a few years later, Hamming's algorithm greatly improved the reliability of data transfer at the time and caused an entire series of improvements that still enable us to trust our computers completely in delivering data intact via the internet to the recipient, despite possible interference.

How can we be sure the data contains no errors?

The simplest way to prevent errors in data storage and transmission is to save the data in more than one identical copy. To increase the reliability of the transmission, the recipient needs to be sent several versions of the data. Yet this simple approach quickly turned out to be unpractical and unreliable, forcing pioneers to discover new ways of increasing the reliability of data transfers. As is frequent in such cases, a source of inspiration was found in an everyday example of language use; an interpersonal communication tool resulting in, for instance, a telephone conversation.

Like telephones, computers must also deal with channel noise and all kinds of line distortions. But the difference is that in a telephone conversation we are usually able to get the message across in spite of the noise. The secret of transmitting information over a bad line lies in the fact that we don't need to understand absolutely everything in order to get the message. The information gets through even if some of it is lost along the way.

Writing numbers with words is a nice example of this trick. If we want to send the number 36172 in writing, but fear that individual numbers might get erased or changed during the transfer, it is more likely that the recipient will get the right figure if we write it out in words: three six one seven two. Even if the words get distorted during the transfer, to read for example "tfree sx ome sevin two", the recipient will still be able to figure it out. Numeric information written in words has a better chance of keeping its form when distorted.

In 1947 Hamming discovered a means of transcribing information for digital computers that was based on the same approach as spelling out numbers. Hamming developed the binary computer code, using only the digits 1 and 0 to enable the computer to detect an error in the transfer and correct it automatically. But this simple method could deal efficiently with only one error at a time. In cases of a number of simultaneous errors, problems were still bound to occur.

A warning is enough

In addition to algorithms for detecting and correcting error, other crucial methods must therefore establish whether information has changed during a transfer. If we suspect an error may have occurred, we simply request a new copy. This is the test used by the majority of internet connections.

The simplest of these distortion detection methods is called checksum. It involves a calculation of the sum of transmitted data, generally encoded as a combination of digits, along with a sum of all the digits that are being transmitted. Practice shows that it suffices to add only the last, decimal digit of the sum in order to confirm the results.

If we add up the digits in the aforementioned number 36172, we get a sum of 19. Then we add the last decimal digit to the existing number, sending the numeric sequence 361729 to the recipient. The information receiver needs to be aware of us using this testing protocol. It enables them to detect a possible error in case of receiving, for instance, the number 361129. The sum of these digits, excluding the last – the error control digit – would in this case add up to 13, meaning that the control digit should be 3 if there were indeed no errors in the transmission. In practice, this kind of information is discarded automatically and the protocol demands that the sequence be sent again.

But even the checksum method isn't completely reliable. A number of errors can occur in the message that rule each other out and result in a sum of wrong digits that matches the one transferred first. This is why error detection methods were eventually upgraded to perform a variety of different computations on the digits, enabling the computer to detect more errors.

The birth of information sciences

Error detection algorithms are only a small part in the large field of information theory. The birth of information sciences is marked by a paper entitled "A Mathematical Theory of Communication", published in 1948 by Claude Shannon. Interestingly, Shannon and Hamming briefly shared an office in Bell's laboratories at that time. They also knew each other from the years both spent working for the Manhattan Project, the research team that developed the first atomic bombs during World War II.

In his groundbreaking paper, dubbed a "Magna Carta for the Information Age", Shannon set up the theoretic pillars of information science. Among other things, he showed that a high-quality data transfer could be achieved despite noise and other interference in the communication channel. He worked out the equations necessary for calculating the fastest possible transfer of information through an unreliable connection.

Without the methods which enable reliable data storage and transmission we wouldn't be living in the information age today, when almost all our crucial data is stored in electronic form. CDs, DVDs, computer disks and other information carriers in mass use today are based on highly perfected algorithms capable of performing auto-detection and auto-correction. The same goes for the central medium of data transfer today; the internet wouldn't have come into being without the extremely reliable algorithms that keep our data safe on its journey through cyberspace.

History

An ancient poem explaining the nature of things

Gian Francesco Poggio Bracciolini was more than an ordinary scribe and emissary serving under popes, cardinals and secular rulers of Italy at the beginning of the fifteenth century. He had another great passion. He travelled around Europe, visiting the monastic libraries of the time in search of valuable ancient manuscripts that might have been misplaced. In 1417, while visiting a German monastery, he had a stroke of luck. By sheer coincidence, he came across a didactic poem of 7400 lines written by the Roman philosopher Titus Lucretius Carus in the first half of the first century BC.

The manuscript, entitled *De Rerum Natura (On the nature of things)* was dated ninth century, which meant that by the time of its discovery, it was already half a millennium old and had probably lain for centuries in the library, stacked at the back of a shelf. If it was occasionally picked up by a monk, he most probably didn't realize that he was looking at a rare testimonial of the then little known but historically important school of epicurean philosophy, or ancient atomists. What Poggio couldn't have realized then, was that his discovery of Lucretius's manuscript would eventually make a crucial contribution to the modernization of Europe and the world.

Poggio's search for lost manuscripts took place before the invention of print, so each text had to be transcribed manually if he wanted to take it with him. Even though he was assisted by a number of scribes, the process was considerably longer than what we are used to today. At the time, Europe was recovering from the medieval plague epidemics that had wiped out a third of the continent's inhabitants, and caused many crafts to be reorganized completely. It was also a time of rediscovery for ancient arts and culture, in which many scholars began looking back to classical civilization's greatest achievements.

Atoms and void

In his long poem, Lucretius describes a Greek school of philosophy founded by Epicurus, following the principles of Leucippus and Democritus. The Epicureans' basic idea was simple but radical, judging by the fact that one can still find arduous opponents to the principles of these Ancient Greek atomists.

The epicureans theorized that nature consisted of two fundamental entities: *atom* and *void*. The world is made of nothing else but undividable atoms of all shapes and empty space between them. Everything is made of atoms, or basic particles, too tiny to be visible to the naked eye, but firm enough for us to feel.

When, for instance, something tastes sour, it is because of its rough and angular atoms, while the taste of something sweet comes from large, round atoms. All our sensations come from atoms mingling with other atoms that ultimately form our bodies and are nothing more than names attributed to certain combinations of particles.

Similarly, the atomists perceived objects as temporary combinations of atoms, no more solid or permanent than dust swirls. Even though they do have a shape, it is only temporary and changes constantly. People are therefore nothing more than a combination of atoms that stick together for a while and finally dissolve.

The radicalism of the ancient atomistic theory isn't only in its simple explanation of the nature of things, but also in the consequences it has for an individual's idea of the world and sense of life purpose. According to some, this radically materialistic cosmic picture implies nihilism and a complete absence of values, on which basis they reject it. Darwinism is the main target of these contemporary warriors for the "preservation of values," since the theory of evolution represents a modern version of ancient materialistic theories.

Mysterious deviation of atoms

The epicurean vision of the world is even more unusual than most people imagine. It envisages more than atoms simply floating around, colliding and forming temporary clusters. The first ancient atomists knew this wouldn't lead to the varied and diverse world we see around us. Atoms are much more than a set of particles moving freely and colliding with each other; they are also completely unpredictable. Every atom is free, which means that it can change course at any time, for no discernible reason. Ancient atoms don't form a large organized mechanism that can be set in motion and then continue ticking like a giant cosmic clock. They are utterly unpredictable fragments that don't abide by any law.

This complete freedom and unpredictability the ancient atomists describe in elementary particles is often left out from the simplified overviews the textbooks give of Epicureanism. But this is precisely what makes atomism so interesting, and for some even socially subversive and dangerous. Lucretius names this freedom *clinamen* or *the swerve* and the term was and still is subject to many interpretations. Among other things, it featured prominently in Karl Marx's doctoral thesis.

It would be wrong to interpret atomic deviation as a result of some mysterious force that occasionally steps in and changes the direction of atoms. It makes more sense to understand this phenomenon in terms of the absence of any rule that would govern the atomic movement. Atoms meander and diverge because there is nothing in their nature that requires them to move in a straight line. If there was such a general rule, it would mean that the atoms are subjected to an external constraint that applies universally. But since the world consists of nothing more than atoms and void, there is no rule on how atoms should move. The swerve is simply a consequence of the fact that the atoms are not required to move in an organized way.

A powerful book that had to be read in secret

In his book *The Swerve: How the World Became Modern* (W. W. Norton & Company, 2011), Stephen Greenblatt observes that Poggio's discovery of Lucretius's manuscript most probably took place at the Benedictine abbey of Fulda in central Germany. Since he wasn't allowed to borrow the book, he had to order a transcription, which took quite some time. But the transcript done by a German monastery scribe proved poor in quality, and Poggio sent it to a friend in Florence who made another transcription in a much clearer and more legible hand. This manuscript was further transcribed and became the basis for later print editions.

Lucretius's poem had an important influence on many Renaissance and new age thinkers. In 1961 analysis of one of the preserved transcripts in the Vatican library showed that the handwriting was none other than that of Niccolo Machiavelli. Because he was very well aware of the controversy surrounding Lucretius's text, which was listed in the manuals of the Inquisition, Machiavelli never mentioned Lucretius's name in his writings. Yet, he studied the work thoroughly and it had an important if unacknowledged influence on him.

In 1989 a heavily annotated printed version of *De Rerum Natura* was sold at auction, and it was later shown that it had belonged to Michel de Montaigne. The great French essayist had added copious notes in Latin and French. Ancient atomism had a great influence on Galileo as well. According to some historians, the ideas in Galileo's work that proved so infuriating to the Vatican stemmed from Lucretius and the theories of ancient atomism.

David Hume - the Newton of moral sciences

In 1734 at the age of 23, David Hume travelled from his native Scotland to France, planning to dedicate himself exclusively to musing and writing. First he visited Paris, but soon found the living expenses in the City of Light to be far beyond his means and so continued on to Reims. He spent a few months there practicing his French, but eventually couldn't afford to live there either, and so moved to a small town in Anjou called La Fleche. The town housed a well-known Jesuit College, which had been attended by René Descartes among other luminaries. The college had an extensive library that Hume made use of for his studies and writings. After three years in France, he returned home with a hefty manuscript that was later published in three volumes as *A Treatise of Human Nature*. Although widely acknowledged in later times as one of the most important books in the history of philosophy, it didn't get much attention on its first appearance. Hume himself joked that it "fell still-born from the press."

Today David Hume is regarded as the most important English-language philosopher of all time, even though in his lifetime he was primarily known as a historian, essayist and political commentator. Still, it was philosophy that intrigued him most over the years, and he kept returning to the ideas he developed in his early revelatory book. He continued publishing philosophical works, trying to bring his views to a wider audience, with practically all his key ideas already formed by prior intensive studies.

The realm of nature and the realm of the spirit

David Hume was born more than three hundred years ago (on May 7, 1711) in Edinburgh, Scotland. He was schooled at home and went on to study at the university, together with his brother, at the age of 11. At 15 he finished his studies but never graduated. He continued to pursue his education autonomously and set himself the goal of becoming the "Newton of the moral sciences". As Newton had demonstrated universal laws in the physical world, Hume set out to discover the laws guiding people and society. This was the main theme of his subsequent philosophical opus.

In the seventeenth and eighteenth centuries, scholars divided themselves between the realm of nature and that of the spirit. The word *moral* had roughly the same meaning as the word *human* and moral sciences were those dealing with human nature and all things spiritual. Today they would be called humanities and social sciences. The domain of moral reasoning included beings capable of cognition, whereas the interpretation of nature was reserved for objects incapable of mental processes.

In interpreting occurrences in the domain of spirituality, scholars based their presumptions on the wishes and will of individuals. A certain act performed by a cognitive being was explained in terms of the entity's desire to perform this act. With non-cognitive objects, their behavior couldn't stem from inner intentions and could thus be explained by means of mechanical triggers.

A characteristic of pre-modern thought was a tendency to interpret occurrences in inanimate nature in the same way used later in describing spirituality. Natural events were seen as direct consequences of the whims and desires of gods and other supernatural beings, guiding the world from behind the scenes.

Descartes tried to limit this "moral" interpretation by dividing the world into a material domain (the realm of extension) and one of mind (the realm of thought), whereby a mechanical interpretation was only applicable to the former. Newton's theory later proved that both the sky and the earth were governed by unified laws, with the exception of a few initial troubles in defining the source of gravity. Cartesians then had a hard time accepting that the source of gravitational force, as described by Newton, remained unexplained in strictly mechanical terms.

Universal law of human nature

Hume took the scientific revolution, which reached its peak with Newton, very seriously. He wanted to expand the scientific approach to nature to the domain of the spiritual or "moral sciences", as the field was then referred to. He searched for the universal laws of "human nature" that govern the spiritual world in the same way as the law of gravity influences the world of material objects. Today we would say he was interested in the functions of the human brain.

In the *Treatise of Human Nature* Hume attempted to formulate the laws governing the succession of our thoughts. How can we tell which thought is reliable and which one is not? The book was subtitled *"An Attempt to Introduce an Experimental Method of Reasoning into Moral Subjects"*. Hume tried to describe the mind in the same way Newton described bodies in space. He wanted to establish a universal law of the mind that was still unconquered by science in the Age of Enlightenment.

He found that the mind is governed mostly by habits or associations. If two events are always observed in the same sequence, we become used to it and start to perceive such a sequence as a whole. This rule controls our mental apparatus in a way that we can't consciously influence. With his "laws of association and habit", Hume came across an important principle of cognition that introduced "mechanical" interpretation into the world of cognitive events, which were previously perceived as mere results of whims and desires.

After giving the question long thought, Hume hypothesized that a universal law of "human nature" rested on the principle of habit. When a sequence of events is repeated, our nature, or rather our brains, spontaneously starts to perceive it as a whole, without any conscious effort on our part. Hume claims the power of habit to be the central principle of human nature.

The idea seems very simple at first, but if one thinks of it universally and in terms of all its possible implications, one realizes that it truly is a revolutionary concept. Stemming from a simple assumption about the functions of human nature and the creation of thought, Hume's analysis caused a complete revolution in understanding the ways of the mind. Immanuel Kant later wrote that by reading Hume's work he was "awakened from his dogmatic sleep" and inspired to start working on his own groundbreaking ideas, which took Hume's to another level.

Too old, too fat, too lazy and too rich

Among other things, Hume's insights offered the conclusion that the "self", the soul, or the central spiritual character that makes us individual, is a mere consequence of the basic mechanical principle he discovered in the human mind. When we perceive ourselves and those around us as consisting of an unalterable essence in which our identities are permanently embedded, regardless of how everything else changes through time, we do so because of the association mechanism that governs our brains.

Hume's insight into the central associative mechanism of human nature is very similar to one that was obtained, many years later and with very different means, by contemporary neuroscience. Clinical research has now established that when individual neurons in the brain are activated simultaneously, a stronger connection is created between them ("cells that fire together, wire together"). This basic principle of the human brain (Hebb's rule) is essentially the same as the universal law of human nature observed nearly three centuries ago by David Hume, although it was then too unconventional to gain any significant following.

After his *Treatise* was published, Hume tried to gain a position as professor of moral philosophy at a number of distinguished Scottish universities, but unfortunately never had any success. His ideas were too radical for him to be allowed to work with students. More problematic still were his sceptical musings on questions of religion, then an extremely delicate issue. His infamous work *Dialogues Concerning Natural Religion* wasn't published until after his death.

Hume worked in a number of different professions during his career. When he was young, he tried his hand as an entrepreneur, a librarian, a diplomat, a secretary, everything but a professor of philosophy. He wrote an extensive and influential *History of England*, which brought him a considerable additional income in his later years. When a publisher asked him to complete an extended edition of his book, Hume replied: "I must decline not only this offer, but all others of a literary nature, for four reasons: Because I'm too old, too fat, too lazy, and too rich."

The artillery of heaven and the heretical rod

On May 10, 1752, the French village of Marly-la-Ville was hit by a storm that would soon enter the history of science. An assistant to the French scientist Thomas-François Dalibard managed to catch some of the clouds' electricity with a metal pole. When the storm approached - the pole was fortunately not hit by lightning – the assistant noticed sparks gathering at the bottom end of the pole, conclusive evidence that thunder storms were linked to electrical occurrences.

In natural science circles of the mid-eighteenth century, the proof of a connection between thunder storms and electricity constituted an important scientific achievement. The sensational news of this discovery quickly spread around Europe. In the weeks that followed, the experiment was tested by other scientists who confirmed the results, while Dalibard immediately sent a report of his findings to the *Académie des Sciences* in Paris, kindly admitting that he had based his experiment on research previously made public by the American polymath Benjamin Franklin.

Flying a thunder storm kite

Contrary to most textbooks on the history of science, Franklin was not the first person to prove, by means of an experiment, that storm clouds contain electrical occurrences. His famous experiment with a kite, which has assumed similar mythic proportions to Galileo's with cannon balls flung from the leaning tower of Pisa, wasn't conducted until June 1752. Franklin didn't know the proof he sought had already been discovered through the experiment in France. Franklin attached a well-isolated metal key to a silken kite and flew it below the stormy skies. When a few minutes later he pulled the kite back down, a jumping spark proved that the key had been electrified by the clouds, offering further conclusive proof of electrical activity in thunder storms.

Even though Franklin's kite experiment wasn't the first to investigate electricity in thunder, it definitely made for a better story. When Franklin's famous political career gained momentum a few years later, the story was taken as gospel and the discovery declared his. Another important reason for Franklin's supremacy in the research of atmospheric electricity was his invention of the lightning rod. In 1811 John Adams, the second president of the United States, in an unexpected litany of praise to Franklin, celebrated his invention of the lighting rod as the greatest recent achievement in science:

> His reputation was more universal than that of Leibnitz or Newton, Frederick or Voltaire, and his character more beloved and esteemed than any or all of them. Newton had astonished perhaps forty or fifty men in Europe; for not more than that number, probably, at any one time had read him and understood him by his discoveries and demonstrations. ... But Franklin's fame was universal. His name was familiar to government and people, to kings, courtiers, nobility, clergy, and philosophers, as well as plebeians, to such a degree that there was scarcely a peasant or a citizen, a valet de chamber, coachman or footman, a lady's chambermaid or a scullion in a kitchen, who was not familiar with it, and who did not consider him as a friend to human kind.

How church-bell ringers got the short end of the stick

Ever since people began building churches on exposed hills, in unpopulated areas in the middle of fields, with belfries towering over the rest of the landscape, they had faced troubles during thunder storms that were difficult to explain. When a storm came raging in, lightning usually struck the belfry, which proved potentially deadly for the worshippers inside. But this wasn't merely a technical problem; it was also theological.

Christians used to believe that thunder storms and lightning bolts were directed by a "higher force" from above, which explained why the lightning was so likely to strike places of worship. Many were convinced that lightning bolts were a sign of God's discontent with the sinful world. The most zealous preachers swore by the power of prayer to keep people safe from lightning, and of ringing the bells to drive the storm away. But the results were not at all encouraging. Bell ringers were high among those professions at the greatest risk of being struck by lightning.

Some religious leaders of the day accused Franklin of interfering with God's authority and claiming the right to explain phenomena that are above human command. When Massachusetts was struck by an earthquake in 1755, pastor Thomas Prince stated in one of his sermons that the frequency of earthquakes may be due to the erection of "iron points invented by the sagacious Mr. Franklin." He argued that "in Boston there are more erected than anywhere else in New England, and Boston seems to be more dreadfully shaken. Oh! There is no getting out of the mighty hand of God."

Battle for authority

By introducing lightning rods to protect church towers, tall buildings and ships, Franklin didn't just enable their owners to save a lot of money, this new technology against natural disasters had an important symbolic meaning. Franklin made a big advance in guiding authority away from traditional religious organizations and redirecting it to newly formed scientific institutions.

In April 1753 he addressed the opponents of lightning rods as follows:

> Surely the Thunder of Heaven is no more supernatural than the Rain, Hail, or Sunshine of Heaven, against the Inconvenience of which we guard by Roofs and Shades without Scruple. But I can now ease the Gentleman of this Apprehension; for by some late Experiments I find, that it is not Lightning from the Clouds that strikes the Earth, but Lightning from the Earth that Strikes the Clouds.

Despite the pastors' criticism, Franklin and his supporters managed to provide some of the exposed American houses, churches and ships with lightning rods. Meanwhile in Europe the idea was met with even greater disapproval. While the Vatican recommended the use of this new technology to protect churches from being struck by lightning, local vicars were wary of lightning rods, believing that they actually attracted lightning. Eventually even the most fanatic opponents of "heretical rods" changed their minds and realized that conductors weren't devil's tricks that infected peoples' minds with unholy ideas.

Lightning is still dangerous

Efficient methods of lightning protection have been known for more than two hundred and fifty years. Yet many houses, mostly in third world countries, are still unequipped with good lightning conductors. Lightning most often strikes in the Tropics, in regions where levels of poverty are high. In 1988, the inhabitants of Cibinong, West Java Indonesia, reportedly witnessed as many as 322 lightning strikes in a single year.

In July 2007 a violent storm hit Ushari Dara, a remote mountain village in north-western Pakistan. After the storm blew over, police reported that lightning had destroyed a dozen houses and killed as many as 30 people. In other less developed parts of the world, lightning still poses a considerable threat to peoples' lives. In 2006 China counted as many as 19,982 natural disasters caused by lightning strikes that took 717 lives and left 640 people wounded. According to the Chinese Meteorological Agency, 109 people died from lightning strikes in August 2007 and 43 were wounded.

The average temperature in China that August amounted to 21.6 degrees Celsius, making it the warmest August since 1951. Higher air temperatures are one of the reasons for unpredictable weather conditions. To alert the citizens to the dangers posed by severe storms, the Chinese Ministry of Education provided 420,000 schools across the country with instructions on lightning strike safety and prevention. The information materials included instructive DVDs, comic books, stories and flyers on how lightning occurs and on how to react if you are caught in a storm.

In poor parts of the world, people often use metal roofing to cover their houses, which makes them ideal targets for lightning. The Bangladeshi government also realized that its citizens should be better informed on the dangers of electrical storms. But since illiteracy in Bangladesh is still very high, the authorities decided to educate the citizens by means of street theatre and popular song lyrics that have a nice ring to the ear and at the same time deliver an instructive message.

A meeting of two minds – one admired, the other feared

In November 1676, a secret encounter took place in The Hague between two of the most influential European thinkers of the day. Gottfried Wilhelm Leibniz was regarded as a rising star of European science and philosophy. His introduction of infinitesimal calculus, which he developed independently of Isaac Newton, became the keystone of modern mathematics. He earned his income as a diplomat and court councilor, devoting his free time to writing illustrious philosophical, mathematical and scientific dissertations.

In The Hague, Leibniz visited Baruch Spinoza, a philosopher and scientist more than ten years his senior, whose reputation was that of the most dangerous thinker in Europe. He was publicly labelled as an "atheist Jew", even though he wasn't really an atheist – rather, his definition of God was extremely unconventional. Ostracized from the Jewish community for his heretical ideas, he made his living polishing lenses by day, and then wrote his philosophical theses by night, undermining as he did so the foundations of society, religion and morality – at least according to his numerous critics.

Leibniz tried very hard to keep his encounter with Spinoza a secret. He was afraid that associating with this notorious heretic, whose works he himself had publicly condemned, would damage his career as a courtier and intellectual. When interrogated later on the subject, he maintained they had met by sheer coincidence while he was in town, and hadn't talked for more than a few hours. And as for their conversation: it was nothing more than a courteous chat about current affairs. But the truth was quite the opposite. The sole purpose of Leibniz's visit to The Hague was to meet with Spinoza. He stayed in town for three days and had a number of long conversations with the presumed heretical mastermind. A short account of their conversations survived, written by Leibniz during one of their meetings, and read aloud to Spinoza. It was a "proof of the existence of God".

Why did "the Courtier and the Heretic" meet in the first place?

The story of the encounter between the two great minds of the seventeenth century is brilliantly described by Matthew Stewart in his book *The Courtier and the Heretic: Leibniz, Spinoza, and the Fate of God in the Modern World* (W. W. Norton & Company, 2006). In addition to an interesting and comprehensive account of their lives and work, the author presents the reasons why Leibniz wanted to see Spinoza in the first place.

Leibniz, nicknamed a "one-man intelligence service", strived for a full overview of not only the political but also the intellectual scene in Europe at the time, while Spinoza, regardless of the heretical nature of his findings, was renowned for his ingenious and influential ideas. While they would not meet in person until a few years later, Leibniz sent Spinoza his first letter on October 3, 1671, addressing Spinoza's expertise on lenses by posing him a few questions about optics, yet making it clear that the correspondence he hoped for would be on the subject of philosophy.

Spinoza was by no means short of other influential pen-friends at the time. When, for instance, Henry Oldenburg, secretary to Royal Society and founder of the first scientific journal was staying in Holland in 1661, he learned about a twenty-eight-year-old philosopher of genius living nearby, who hadn't yet been able his work published. Oldenburg decided to pay him a visit and subsequently started a correspondence with Spinoza. They met in the garden in front of Spinoza's house and talked for several hours.

An outcast from the outcasts

Spinoza was a descendant of Portuguese Jews who migrated to Holland in search of refuge from the Spanish inquisition. The Dutch local authorities didn't bother them, but on condition that they wouldn't spread heretical ideas or cause any other trouble. As it turned out, this unofficial agreement posed a great problem for the young philosopher. Spinoza favored the new strictly rational approach to philosophy that had been introduced recently by René Descartes. When rumors started to spread of Spinoza defending views that were considered morally offensive, he was confronted by a group of Jewish leaders who warned him to revoke his statements publicly or else face serious consequences.

Spinoza refused to cooperate with the Jewish patriarchs, who punished him by excommunication. This meant that no other member of the Jewish community, including his family, was allowed to have any contact with him. To them, he was as good as dead. If anyone met or talked to him, they too would face the same punishment. Spinoza spent twenty years in forced isolation, at least as far as his Jewish family and friends were concerned. But this didn't stop him from developing his nonconformist ideas further. His expulsion in 1656 marked an important turning point in his life: he ceased to belong to the Jewish community, yet in turn became a free citizen of the world.

Spinoza died only a few months after his meeting with Leibniz. His premature death at the age of forty-four was reportedly due to lung illness, a result of breathing in glass dust from the lenses he ground.

Radical Enlightenment

The *Ethics*, Spinoza's best known work, in which he described his views of the world in the form of axioms, theorems and derivations, was published only after his death. More controversial in his lifetime was Spinoza's analysis of the Bible and other religious and political texts, grouped together as his *Theological-Political Treatise.*

In it, Spinoza showed that the Bible shouldn't be interpreted literally as the written Word of God, but as the work of actual people. He convincingly proved that the truths contained in the Bible weren't scientific facts, but rather a collection of moral allegories that are meant to provide us with guidance on how to live our lives rather than explaining the world as it is.

This interpretation is widely recognized today, but in Spinoza's time it was nothing short of revolutionary and therefore highly dangerous. Spinoza's radicalism was evident not only in his criticism of religion and the governing ideology of the time, which established secular rule on the basis of religion, but also in introducing rational thought as a pillar of equality supporting all people.

All people are equally entitled to the liberty of freely applying their reason, argued Spinoza. We are all capable of thinking and thought is our universal right. Our customs and traditions, values, language, upbringing and other cultural traits are tied to the specific nurturing environment and cultural tradition an individual belongs to, and therefore are not universal. The ability to apply rational thought, however, is something available to everyone. The only difference is that some people make use of it more frequently than others.

The ability of all people to apply their reason, and thus stand as an equal member of the community, is the key principle of the Enlightenment. Some historians place Spinoza at the source of the most radical version of this historical movement. Enlightened absolutism and other uncompromising interpretations of the Enlightenment eventually gave rise to the modern concept of the democratic state, in which all people are entitled to universal human rights, freedom of thought and religious belief.

The concept of radical egalitarianism among people who accept the principles of argumentative public debate quickly established itself as the foundation of the scientific community, in which the content of an argument is more important than the person who utters it. Anyone who abides by the rules of the rational exchange of thought is welcome to contribute their ideas, which are later assessed by the scientific community. Unfortunately the principles of enlightenment in politics and wider society didn't work out as well as in science. One important reason for this is that even though all people are capable of applying their reason and basing their decisions on relational analyses, in practice we don't do this as often as we could. We tend to be guided automatically by the habits, beliefs and values of the cultural environment we live in. To stay alert and rationalize everything that is going on around us is not only very difficult but even practically impossible.

The magnificence of Chinese science revealed

It is often the case with great names in the history of science that they make their most important discoveries through a pastime, rather than while doing their regular job. This was certainly true of Joseph Needham, the English biochemist and Cambridge professor. He was a brilliant scientist, awarded a fellowship of the Royal Society for his work in biochemistry, but today he is remembered for something altogether different.

A few years before World War II, when he was already a well-established scientist, Needham fell in love with a Chinese researcher who came to train in his laboratory. He was later enchanted with China itself, a land he studied intensively for many years until his death in 1995. He wrote a series of books on the history of Chinese science and technology that offered the Western world an entirely different view of the past achievements of this distant and always mysterious civilization.

A passionate smoker who lived almost a hundred years

Science wasn't Needham's only talent. He also had a flair for languages. As a teenager he mastered eleven languages, including Latin and Greek, and had other strong interests. A devoted Christian and political leftist, he never entered the Communist party but remained a convinced socialist, publicly proclaiming his views and opinions. He was a passionate smoker, nudist and, according to his biographer Simon Winchester, a notorious womanizer.

Needham was married for more than sixty years to an equally successful scientist, an expert in the biochemistry of muscles, Dorothy M. Needham, who was also a member of the Royal Society. Yet the couple shared somewhat liberal views on marriage. When in 1936 a young Chinese researcher, Lu Gwei-djen, joined them at Cambridge, all three became good friends and spent a lot of time together outside the laboratory. After a few months, Lu Gwei-djen and Needham started a romantic relationship, with his wife stoically observing the romance without protest when the lovers made no attempt to hide their affair.

This new love triggered a new interest in Needham, a fascination with the Chinese language. After he and Gwei-djen first slept together, he lit a cigarette and asked how it was pronounced in Chinese. Cigarette was the first logogram he learned. He went on to dedicate most of his free time to reading and writing in Chinese. Thanks to his rare linguistic talent he made swift progress, and by 1938 was fluent in reading and writing the language.

Saving Chinese science

By the end of 1939 news of the disastrous state of Chinese universities and institutes reached England. Then, during the war, those Chinese scientists who managed to travel from the Far East to Europe reported how Chinese universities were being systematically eradicated by the Japanese. University buildings were first bombarded and later completely destroyed by ground forces.

In November 1939, a group of leading scholars from Oxford and Cambridge gathered at the home of a distinguished professor of eastern religions to meet a young Chinese academic who was able to brief them on the actual state of Chinese science. On learning of the terrible conditions in Chinese universities, British scholars agreed it was England's duty to help. They found it of utmost importance to send a delegate to act as a relations officer and rescue coordinator.

They initially intended to appoint an eminent Oxford professor of Chinese philosophy to the position, but when Needham came to mind, everyone agreed he was the right person for this difficult task. He spoke a number of languages, including

Chinese, and with his vast knowledge of natural sciences he could easily discuss the chemicals and laboratory equipment that might be needed. Most importantly, he was infinitely fascinated with China.

When Needham learned of his appointment, he was thrilled. He would finally get to know the land to which he had been drawn for so long. But it still took years of getting through British red tape before he was finally allowed to make the journey.

Travelling around China

Needham set out for China towards the end of 1942. He began by sailing on a cargo ship for several months to India, and flew from there to China in March 1943. He landed in a small provincial town and immediately made contacts among local scientists. While learning of the conditions facing Chinese science, he became familiar with the fascinating history of Chinese discoveries and technological inventions.

By the end of the war, he had travelled across the entire country and visited even the most secluded institutes, offering help to their researchers. He proceeded to order the chemicals, lab equipment and literature from England that were most needed. During his mission, several tons of scientific supplies were shipped over to China, providing scientists with components that were crucial in saving Chinese science from vanishing completely during the war.

But what proved even more important than direct material help was Needham's meticulously collected information on the history of Chinese science. Wherever he stayed, he tried to get as much information and literature on Chinese history as he could. He bought old books in remote parts of the country then sent them home to Cambridge, having resolved to write a book on the history of Chinese science.

When he returned home after the war, he immediately got to work. Initially he planned to write a single concise work, but by the time he died, Needham's encyclopedic *Science and Civilisation in China* had reached 17 substantial volumes. The work is still being continued by his successors.

Needham discovered that China was the origin of many important inventions, such as gunpowder, print, toilet paper and the compass. But despite its past achievements, Chinese science came to an almost complete standstill during the last few centuries. Needham couldn't understand why Chinese research and knowledge hadn't evolved as it had in Europe, even though it seems that all the necessary conditions existed. "Needham's question" still remains unanswered today. Sinologists more or less agree that development in China was slow because of its society's lack of need for constant improvement and innovation, but there is no consensus on why this was the case.

Biological weapons and a lecture on gunpowder

After the Korean War, Needham was invited by friends from China to carry out an investigation with fellow experts on whether the American army had used biological weapons during the war. They noticed an increase in cases of anthrax and similar diseases that could be linked to the use of such weaponry. After a few months of research, Needham's commission, which included a number of scientists, established that Chinese accusations were justified. As a result, Needham faced public accusations upon returning to his homeland that amounted to media lynching. He nearly lost his university position, and his reputation was severely damaged.

He was barred from entering the United States for twenty-five years and when he was finally granted a US lecture tour at the end of 1970s, another bizarre incident occurred. Needham held a public lecture in Chicago on the Chinese origins of gunpowder.

Among the people sitting in the audience was an eccentric mathematician called Ted Kaczynski, whose essay on the evils of contemporary civilization had recently been rejected for publication. Listening attentively to the lecture, Kaczynski learned about an ancient Chinese design for a simple exploding device. Six weeks later, the professor who had rejected Kaczynski's essay received a wooden crate that exploded in the hands of a security guard at the university, leaving the man with significant injuries. Over the following two decades, Kaczynski sent out numerous letter bombs, killing three people and wounding twenty. He was nicknamed the Unabomber by the American media, and wasn't captured until the FBI tracked him down in a secluded mountain area in 1996.

When Needham's book on the history of Chinese science attained major success, his public reputation was restored and his achievements recognized in the Queen's 1992 Honors List. Needham's wife Dorothy died in 1987, having reached a venerable age and witnessed the opening of the institute devoted to the study of China's scientific history that now bears Needham's name. Two years after her death, Needham married Lu Gwei-djen, who had been waiting 51 years for that moment. When they died, their ashes scattered together in the garden of Needham Research Institute in Cambridge.

The "traitorous eight" make it on their own

In June 1957, eight brilliant young scientists held a secret meeting at Clift, a luxury hotel in San Francisco. About a year and a half earlier, they had moved to the West Coast with their families to work at the Shockley Semiconductor Laboratory, a research facility founded by the co-inventor of the transistor, William Shockley. But despite their initial enthusiasm, the young scientists were far from satisfied with their new employer. His increasing authoritarianism was getting in the way of their original ideas and inventions, so they decided to leave the lab and establish their own company, led by their twenty-nine-year-old colleague Robert Noyce.

Their break from the company was considered an extraordinarily brave thing to do at the time. Despite their family responsibilities the young experts decided to leave their solid positions in the hope of gaining new liberties that would enable them to do what they truly believed in. They formed new values in their own company, breaking with the tradition of a strong hierarchy in American corporate culture. For them, the result of one's work was more important than the working hours, and ground-breaking ideas and innovations were highly appreciated.

Since the "traitorous eight" – as the group was later named – didn't bring any paper to their secret meeting, they validated their newly formed company by signing a number of dollar bills. This was the world's first "spin-off", a newly formed company created by a person or a group within a parent organization. This new kind of development eventually formed a model for much later entrepreneurship, still largely evident in Silicon Valley.

The beginnings of electronics

The first electronic devices produced after World War II were large and wasted a lot of energy. They consisted of vacuum tube components that looked like oblong light bulbs. They had a short lifespan and required a great deal of maintenance. That is why the invention of the transistor, the vacuum tube's much smaller, more reliable and more efficient counterpart, presented such an important technological achievement. The first commercially available electronic transistor-based device was a small pocket radio receiver made in 1954, which was referred to simply as "the transistor".

One of the key figures in the invention of the transistor at Bell Lab was none other than William Shockley. Soon after his initial success, he decided to found his own electronics research company. He gathered talented young scientists and engineers from all over the world and brought them to California, to the place that is now called Silicon Valley.

Before William Shockley transferred his development lab facilities to the Californian valley, which was then still covered in orchards, the most prominent high-tech companies established their bases along the US East Coast. Shockley was lured to the west by family ties, as well as the low-cost land offered to developing companies by Stanford University. The first goal that Shockley Semiconductor Laboratory set out to achieve was to improve their production technology so that transistors could function in less ideal or more extreme circumstances. This meant shifting the material that semiconductors were made out of from germanium to the more stable silicon.

About six months after the precocious young scientists gathered in California, they got the news that Shockley, their boss, along with his two colleagues, had received the Nobel Prize for Physics for inventing the transistor. Normally this sort of news would have had a stimulating effect on the team, but unfortunately it caused nothing but conflict. Shockley, known for his arrogance and narcissism, became downright impossible. He behaved more autocratically than ever, showed signs of paranoia, repeatedly made bad decisions without consulting colleagues and later vented his frustration on them when things went wrong.

Seven young experts refused to work in the oppressive atmosphere and reported the problem to the owner of the company, who made it clear that he wasn't ready to depose a Nobel Prize-winner just like that. The scientists realized they had no other option but to resign. This was practically unheard of in America's corporate culture at that time. Upon arriving in California, the young engineers were convinced they would be working for the company until they retired. It was considered an unwritten rule back then, whereas today it goes almost without saying that employees may leave their work at any time to pursue their own ideas.

When the defectors considered what to do, they were advised to open their own development business and find an investor who would recognize the commercial value of their work. Since they were all scientists rather than managers, they realized that if they wanted their enterprise to succeed, they would need to convince Robert Noyce to join them. He was an excellent engineer and incidentally also a great leader.

This is why the eight of them met in the Clift Hotel the summer of 1957 and decided to make it on their own. They made their story part of our collective history not only by upgrading the transistor and creating almost a hundred new spin-offs, but, under Noyce's leadership, inventing and launching the world's first microprocessor, the foundation of nearly all the electronic devices in use today.

The invention of the microprocessor and the beginning of the computer era

When in October 1957 the Russians successfully launched Sputnik, their first artificial satellite, the Americans were shocked to find themselves lagging behind their competitor. The American Government responded by founding NASA, and by heavily subsidizing technological research. Their defense industry required advanced transistors for use in aviation and aeronautics and Noyce was able to generate good business in those fields; more problematically, he also promised to supply the Department of Defense with a transistor which didn't yet exist. Still, his team of brilliant young engineers rose to the challenge and successfully delivered the order in just five months.

But the transistors of the age were still not reliable enough to meet the demands of the military. A new foolproof design had to be found. There was also a need for wiring multiple transistors into a one-piece unit, because electronic devices were getting increasingly complicated and required circuits which linked various electronic components. In March 1959, Jack Kilby developed a method of creating a complex electronic circuit in a single chip for the company Texas Instruments. Noyce and his colleagues got to work immediately, trying to develop their idea for an integrated circuit, not by connecting individual components on a silicon panel with tiny wires, as Kilby suggested, but rather printing links directly onto the carbon-like slice. This was the birth of the world's first microchip, the basis of the modern electronic industry.

As is often the case with new discoveries, at the end of the 1950s two scientists from two different backgrounds came up with an almost identical idea. Kilby was the first to publish his findings, while Noyce came up with a production design that proved more efficient. After years of patent war, the two companies finally reached an understanding and agreed to share the microchip production license. Despite their success in making increasingly cheaper and more efficient microchips over the years, Noyce, as the Fairchild Semiconductor's CEO, refused to share the company's profits with the board of trustees, accusing them of not investing enough into the development of the company and giving stingy awards to its most talented researchers, who were continuously leaving the company for better positions.

That is why in 1968 he decided to co-found another company, together with Gordon Moore, author of the famous eponymous principle that the number of transistors on integrated circuits doubles approximately every two years. Moore's Law stands as both an observation on progress in electronics, and a challenge to the industry. Noyce and Moore named their company *Intel.* They invited Andy Grove, a talented young engineer of Hungarian background, to join their team. Grove became the company's director of engineering and later a very successful CEO – it was under his leadership that the firm developed into the world's greatest semiconductor and processor-producing company.

When he founded Intel, Robert Noyce was 40 years old. In the course of his subsequent career, he supervised the invention of the first microprocessor or computer chip, the core of the majority of complex electronic devices we now use in everyday life. Because of his decisive influence as a director and expert in introducing new values to high-tech business management, he was nicknamed "the Mayor of Silicon Valley".

Babies, between Kant and the Molluscs

In *A History of Western Philosophy*, Bertrand Russell wrote that from the seventeenth century on, "almost every serious intellectual advance had had to begin with an attack on some Aristotelian doctrine". Western thought remained glutted with Aristotle's theses and theories for more than a thousand years, which consequently meant that any new theory was primarily defined by the criticism it provoked towards this giant of ancient philosophy.

A similar situation to that which Russell described arose in the field of developmental child psychology in the second half of the twentieth century. From the 1960s to the 1980s this scientific field was dominated by the theories of the Swiss-born psychologist, scientist and philosopher Jean Piaget. Any progress in scientific research on mental development in children was necessarily derived from Piaget's theories, in an attempt to complement or upgrade them.

His first scientific paper, written at the age of ten

It may seem strange, but the most prominent child psychologist never received a formal education in psychology. Neither did he enjoy a typical childhood. Piaget's father was an authoritarian man and professor of history, who specialized in The Middle Ages. Jean's mother was mentally unstable and was repeatedly hospitalized. Thus, the Piagets' family life was far from idyllic. In his childhood years, Jean enjoyed studying birds, fossils and shells, as well as building mechanical devices, anything to get his mind off the oppressive family environment. He wrote a book on the birds of his home region in Switzerland before the age of ten.

But Piaget's father instantly clipped his son's wings by commenting that the book was merely a compilation of different sources and not an original work. Piaget therefore decided at ten years old that he would make sure no one had any grounds for reproaching his future research. When he spotted an albino sparrow in the park, he wrote a short scientific report and sent it to the local natural sciences magazine. The editor published it in 1907 without any remarks, without realizing that its author was in fact a ten-year-old boy.

Encouraged by the article's success, Piaget wrote to the director of the museum of natural sciences in Neuchâtel, his hometown, asking for access to the museum's collection. The director granted the young scholar's request and allowed him to classify and catalogue the shell collection. Piaget spent four years working in the museum for four hours, twice a week, and eventually acquired enough knowledge to start publishing articles on molluscs in zoology magazines before the age of sixteen.

Of course none of the readers knew that the author of these scientific texts was so young. Communication in academic circles was still very slow in those days and scientists across Europe presumed that the person behind the increasingly distinguished name was a young postdoctoral expert on molluscs, rather than an adolescent schoolboy. An amusing event occurred when the director of the central Swiss museum of natural sciences in Geneva sent a letter to Piaget, inviting him, solely on the basis of his publications, to curate the museum's newly acquired molluscs collection. Piaget politely refused, since he still needed to graduate from high school at the time.

How babies think

Piaget's interest in zoology was interrupted by a long vacation with his godfather, who introduced him to other scientific fields, especially philosophy. The young mollusc expert was immediately drawn to philosophical questions, of which he was most fascinated with epistemology. By the end of the holidays, he had a new goal in life: explaining the development of knowledge from a biological point of view with an emphasis on how reliable knowledge is acquired. "Piaget wanted to find a link between Kant and the molluscs, between epistemology and biology," is how his scientific project was tellingly described by one biographer.

Piaget's key idea was to reach the final goal of his research (explaining the process of knowledge acquisition) by systematically studying child development from the earliest age. He began by making notes of everything that went on in the lives of his three children. Jacqueline, Lucienne and Laurent Piaget became the main figures in their father's books and scientific papers.

By reading his work, we can learn more intimate details about Piaget's children than we may ever observe in our own. His "home laboratory diaries" describe every minute of every event and change in his children, including some that would most probably pass unnoticed by other parents. Moreover, it is important to note that his observations took place in the 1930s, without the use of any video or sound recording equipment, only pen and paper. The other person deservingly credited for the amount of the information collected was Piaget's wife Valentine, whom he met at university and who was initially his student.

Children as a foreign species

Piaget was never interested in the conventional questions in child development posed by the majority of his contemporaries. He tried to approach problems from a different angle. For instance, he wasn't interested at what age children develop a certain form of knowledge or ability, but rather why all children, until they reach a certain age, make the same mistakes trying to acquire this new ability.

Piaget was certainly not the first researcher who looked upon children as if they were a foreign species. Darwin was at least one other such example. But Piaget was the first to treat babies and children not as simplified versions of adults, but as creatures whose ways of functioning and perceiving the world greatly depend on their current stage of development.

Piaget's observations and experiments showed that babies' experience of the world is deeply mysterious and fairy-tale like. He noticed that until the age of about nine months, a baby doesn't have any awareness of the duration of things. They can become so fascinated with a toy that they start crawling towards it immediately, but suddenly lose their interest if we hide the toy by covering it up with a cloth, for instance. If we cover the toy with something transparent and the child can still see it, they won't get distracted at all.

Nowadays it probably wouldn't be easy to find someone who believed an object's existence depended on whether it is visible or not. Maybe there is an eccentric philosopher somewhere musing on such ideas, but in everyday life, one couldn't function properly by believing things don't exist if one doesn't see them. However, Piaget's research showed that, up to a certain stage in their development, babies believe just that. Objects appear and disappear as if everything was part of a fairy tale or magic trick.

The image of the world grows with us

To put it simply, Piaget's theory states that a child's cognition develops in certain phases that cannot be compared to simplified versions of adult cognition, but consist of specific characteristics for each phase of development. Similarly to the way that the mental categories through which we understand the world around us are heavily dependent on the cultural environment we are brought up in, children's cognition depends on their current stage of development. Overall, Piaget's work is understood as having founded a "constructivist" approach to learning, which looks for the links between our experiences and our ideas as the source of what we know.

Over the last few decades, the development of new scientific approaches has enabled researchers to observe specific brain activities in real time, leading to a series of new findings that provide a significant upgrade on Piaget's classical theory. Yet such advances still share the same mission that the young expert in molluscs first set himself before embarking on his life's real work. He opened our understanding of the children we once were, and the children we bring up. Like many major figures in the history of science Piaget lacked a childhood, but not the child-like qualities of curiosity and imagination that are essential to any great scientist.

A NOTE ON THE AUTHOR

Sašo Dolenc (*saso-dolenc.com*) is a physicist and philosopher by training. He is a co-founder and the editor in chief of the popular online science journal Kvarkadabra (*www.kvarkadabra.net*), whose mission is to explain science in a simple, attractive and fun form that is open to all.

The aim of the articles collected in the series of *Short Stories from Science, History and Philosophy* is to present some key scientific events, ideas and personalities in the form of short tales that are easy and enjoyable to read. Scientific and philosophical concepts are explained in a way that anyone may understand. Each story may be read separately, but at the same time they all band together to form a wide-ranging introduction to the history of science and areas of contemporary scientific research, as well as some of the recurring problems science has encountered in history and the philosophical dilemmas it raises today.

In Praise of the Series

"Each chapter in this book explains a complex problem through moving, amusing and marvellous stories. Sašo Dolenc's recurring theme is the elusive and often eccentric nature of inspiration; but in exploring it he covers an immense variety of subjects, from meteorology to microbiology, computer technology to market theory. His readers will gain a succinct and satisfying lesson on each topic, and a sense overall of the labour, genius and luck that science demands."

– **John Stubbs**, author of *John Donne: The Reformed Soul* and *Reprobates: The Cavaliers of the English Civil War*

"Great fun. Like Malcolm Gladwell, Dolenc writes about complicated science in a clear, accessible way that entertains and educates. The smarter and better the writer, the clearer and simpler he will make concepts that are difficult to grasp. Reading this book is a pleasure you can learn from."

– **Noah Charney**, best-selling author of *The Art Thief* and *Stealing the Mystic Lamb*

"If I were the only survivor on a remote island and all I had with me were this book, a Swiss army knife and a bottle, I would throw the bottle into the sea with the note: 'Don't worry, I have everything I need.'"

– Ciril Horjak, alias **Dr Horowitz,** a comic artist

"Does science think? Heidegger once answered this question with a decisive No. The writings on modern science skillfully penned by Sašo Dolenc, these small stories about big stories, quickly convince us that the contrary is true. Not only does science think in hundreds of unexpected ways, its intellectual challenges and insights are an inexhaustible source of inspiration and entertainment. The clarity of thought and the lucidity of its style make this book accessible to anyone … in the finest tradition of popularizing science, its achievements, dilemmas and predicaments."

– **Mladen Dolar**, philosopher and
author of *A Voice and Nothing More*

"Sašo Dolenc is undoubtedly one of our most successful authors in the field of popular science, possessing the ability to explain complex scientific achievements to a broader audience in a clear and captivating way while remaining precise and scientific. His collection of articles is of particular importance because it encompasses all areas of modern science in an unassuming, almost light-hearted manner."

– **Boštjan Žekš**, physicist and former president
of the *Slovenian Academy of Sciences and Arts*

"The writing is understandable, but never simplistic. Instructive, but never patronizing. Straightforward, but never trivial. In-depth, but never too intense."

– **Ali Žerdin**, features editor at *Delo,*
the main Slovenian newspaper

Made in the USA
Monee, IL
05 November 2026

67811372R00125